THE SUBURBAN MICRO-FARM

The Suburban Micro-Farm: Modern Solutions for Busy People

Printed in the United States of America
Originally published in 2017. First full color edition printed in 2018.

Twisted Creek Press
1081 St. Rt. 28, #256
Milford, OH 45150
TwistedCreekPress.com

Designers: MelindaMartin.me, Happenstance Type-O-Rama
Editors: Julie Hotchkiss and Brannan Sirratt
Illustrator: Becky Bayne
Indexer: Lydia Jones

ISBN: 978-0-9975208-3-5
Library of Congress Control Number: 2017916241

Publisher's Cataloging-In-Publication Data
(Prepared by The Donohue Group, Inc.)

Names: Stross, Amy. | Bayne, Becky, illustrator.
Title: The suburban micro-farm : modern solutions for busy people / Amy Stross ; illustrator: Becky Bayne.
Description: Full color edition. | Milford, OH : Twisted Creek Press, [2018]. | Originally published in 2017. | Includes bibliographical references and index.
Identifiers: ISBN 978-0-9975208-3-5 (print) | ISBN 978-0-9975208-4-2 (ebook)
Subjects: LCSH: Vegetable gardening--United States. | Farms, Small. | Organic gardening. | Edible landscaping. | Permaculture.
Classification: LCC SB321 .S87 2018 (print) | LCC SB321 (ebook) | DDC 635--dc23

THE SUBURBAN MICRO-FARM

MODERN SOLUTIONS
FOR BUSY PEOPLE

AMY STROSS

For Vince,
who always believes in me:
It is my joy to spend this lifetime with you.

And for everyone who
started with a seedling and had the courage to plant it.

CONTENTS

Bonus Materials: *TenthAcreFarm.com/tsmf-companion*

LIFE BEGINS
THE DAY YOU
START A GARDEN.

CHINESE PROVERB

INTRODUCTION

The word *farm* evokes images of rolling hills, pastures of grazing livestock, extensive rows of crops, and old barns. Thus, it's tempting to think that *micro-farming* in the suburbs is simply a miniature imitation of that rural subsistence life, sought by those who fantasize of country life but can't leave the city for some reason or another.

While some suburbanites may dream of one day owning rural land, many more are firmly rooted in their convenient, suburban location that affords optimal opportunities for work, school, entertainment, and amenities. Yet in this unlikely land of sprawling lawn, we are seeing a rising wave of suburbanites growing food or keeping livestock in their yards.

"The times they are a-changin'," to quote Bob Dylan, the quintessential balladeer of cultural shifts. As the number of professional farmers—as well as farmland—continues to shrink in modern times, suburbanites are picking up their digging forks and transforming those lawns into productive spaces. The face of modern farming is becoming more diverse, and it couldn't have come at a better time. Questions of food security are arising in developed nations that depend on long-distance supply chains and lack contingency plans in the face of crisis. In this realization, one thing has become clear: our local environments are underutilized and underappreciated.

But what could be the appeal of farming in such civilized suburban digs, whose focus until now has been strictly recreation and leisure? As Bill Mollison, the father of permaculture, puts it: "The only ethical decision is to take responsibility for our own existence and that of our children." In short, suburbanites have decided to take an active role in providing sustenance for their families, and they've committed to do so even with busy schedules and—sometimes—micro-sized spaces.

Indeed, while modern, suburban micro-farms may take on a different look than their traditional farming forebears, the definition of a farm remains true: "an area of land used for growing crops or keeping livestock, typically under the control of one owner." Therefore, homeowners who want to create a more productive landscape—whatever its method or appearance—can be called farmers!

In this book, I'll share techniques to make the suburban landscape manageable for the modern, part-time micro-farmer. I'll use the word *suburban* loosely, for the techniques and strategies in this book could certainly be used in any small space, whether urban, suburban, or rural. The suburban concept casts a wide net, with many micro-farmers identifying themselves as living in semi-urban or semi-rural spaces. As we'll see throughout this book, even those with large properties may wish to design a compact micro-farm so that it is easily manageable. There are even tips for those with no land at all.

One example of a management strategy for rural, large-scale farms is row cropping. However, this style may not have the aesthetic appeal that a suburban homeowner prefers, nor be appropriate for the hand-tended farm. In this book, we'll take a look at ecologically friendly food production practices that are suitable for residential landscapes.

While some micro-farmers will be drawn to skills such as practicing fiber arts or keeping livestock like chickens or bees, for example, this book will focus on growing fruits, vegetables, and herbs. You will get tools to help you start and maintain a micro-farm in your suburban yard so that you feel empowered and motivated to care for the garden and enjoy the harvest. I'll share the techniques I used to make my own suburban landscape manageable, including tips for creating an edible front yard, and how to get along with neighbors who can't make sense of your farming pursuits.

While there are a gazillion books and resources out there about gardening, very few of them address the issue of time. It's difficult to imagine finding the time to keep gardens and orchards, cook fresh meals and preserve the excess, all while maintaining other aspects of modern life. Once we admit that we are all busy, we can seek out processes to help us produce food on a limited schedule. You'll get my bare-bones life hacks for growing a productive garden while still tending to all of life's other calls. These will help to ensure you don't develop what I call *Garden Overwhelm Syndrome*.

WHAT IS PERMACULTURE?

Permaculture is a system for designing agricultural landscapes that work with nature instead of against it. I like to call it *edible restoration,* since the tools used in permaculture can help to restore land as well as yield food for humans. Permaculture design can also be used in myriad ways to improve the efficiency of other systems, such as city planning, energy, waste, health care, and so forth. In this book, however, I will focus on permaculture as an approach for growing food efficiently with ecological integrity. While it is implicitly weaved throughout the book, see chapter 11 for more permaculture-specific tools.

This book will not only teach you the basics of growing food, but it will also show you how to organize your busy life around your micro-farming activities. We can look at pictures of beautiful gardens all day, but actually having a plan, getting started, staying on track, and using time efficiently are paramount to being successful micro-farmers in the long term.

"A goal without a plan is just a wish," according to the French writer Antoine de Saint-Exupery. As such, at the core of this book is my strategy for developing a plan, prioritizing tasks, and keeping records. I'll share my process for organizing my micro-farm—along with a host of supplemental materials—to help you confidently turn your dream of a productive yard into a reality. The downloadable, supplemental materials will guide you in planning what to plant and when, utilizing checklists and monthly calendars to stay on track, and keeping records.

The no-nonsense how-to sections of this book will help you tackle all kinds of projects for setting up a micro-farm and maintaining it throughout the season. While some of this information can be found in other gardening books and resources, those sources usually assume that the farmer has an ideal, flat, sunny piece of land. This book will show you how to deal with less-than-ideal landscapes, such as utilizing the front yard tastefully, from small spaces to no land at all; and from shady areas to sloping land.

Digging deeper, I'll even share permaculture strategies that can help you create an ecologically friendly micro-farm while reducing your workload. By modeling nature, permaculture will help you take your garden to the next level.

To sum it up, this book will help busy gardeners learn how to work with the land they have—with the time they have available—to create a beautiful and ecologically friendly micro-farm.

If you have dreamed of growing food but the demands of modern life have left you feeling overwhelmed, then this book is for you.

GOOD FARMERS, WHO TAKE SERIOUSLY THEIR DUTIES AS STEWARDS OF CREATION AND OF THEIR LAND'S INHERITORS, CONTRIBUTE TO THE WELFARE OF SOCIETY IN MORE WAYS THAN SOCIETY USUALLY ACKNOWLEDGES, OR EVEN KNOWS.

THESE FARMERS PRODUCE VALUABLE GOODS, OF COURSE; BUT THEY ALSO CONSERVE SOIL, THEY CONSERVE WATER, THEY CONSERVE WILDLIFE, THEY CONSERVE OPEN SPACE, THEY CONSERVE SCENERY.

WENDELL BERRY

PART I:

GETTING TO KNOW THE MICRO-FARM

1 SUBURBIA: CIVILIZATION'S OPPORTUNITY

FROM A SMALL SEED, A MIGHTY TRUNK MAY GROW.

AESCHYLUS

THE SUBURBAN MICRO-FARM

At age 33, I quit my job as a high school teacher. Although I enjoyed teaching, the formal education environment didn't quite suit my personality. I was drawn to teaching and learning through independent, project-based adventures, and I longed to be outside in the fresh air and warm sun, away from a bell schedule and fluorescent lights.

As anyone who has started down a career path only to discover it wasn't a good fit might understand, the mismatch between my job and me took a toll on my well-being. I quit after a particularly grueling year. The decision wasn't typical for my driven, type-A personality, yet I felt helpless in the wake of worsening health.

That summer, I moved to a new city to be with my then fiancé. It wasn't an ideal way for us to begin a life together, to say the least! The financial burden of starting out as a single-income household wasn't the heaviest weight on our early relationship, however. Rather, it was the deep sorrow in my heart for not having found a way to be useful to the world.

YOU CAN BURY A LOT OF TROUBLES DIGGING IN THE DIRT.

AUTHOR UNKNOWN

I wandered around a bit, trying to figure out how to be a contributing member of my household. In the process, I worked several part time jobs, one of them as a gardener with a residential gardening company. The "dirt therapy" helped me work out some of my heartache, allowing me to focus on improving my health, as well as the joy of planning a wedding.

I had never gardened before, and the job was exactly the hook that I needed. Sun, fresh air, exercise, and thought-provoking design challenges all improved my mental state and confidence. In fact, to be more accurate, the experience changed my life. I couldn't learn enough about growing things. I went on to take various gardening classes at my local garden education organization, becoming certified in both community garden development and permaculture design.

From that point on, I felt like I had discovered a way to contribute to our household that was both useful and brought me joy: growing food! Mr. Weekend Warrior (the nickname for my husband that I'll use throughout this book) and I started small by joining a CSA (community supported agriculture). Getting fresh produce each week from a local farm and figuring out what to do with it all was a great way to start. I can't imagine how overwhelmed I would have been if I were growing all of our own produce for the first time while also trying to figure out how to use and preserve it in the kitchen.

We also installed rain barrels and a rain garden to irrigate our ornamental landscaping without using municipal water. I integrated vegetable crops into the existing landscape. At that time, I assumed I needed a lot of land to grow any decent amount of food—like the vast fields of our local farmer—so I didn't aspire to give my micro-sized yard an edible makeover. Instead, I started a landshare project, in which I drew up a legal contract and grew crops on someone else's semi-rural property in exchange for half the harvest. By the end of the first season, however, I realized that I could have grown the same amount of produce right in my own small backyard. It was an *aha* moment.

I decided I should focus on my own home.

With Mr. Weekend Warrior on board with my crazy plan, we added more gardens and increased food production each year in our own yard as we learned and mastered skills. Years later, I now look out with awe and pride at our edible front yard, which captures rainwater from the roof and includes cherry trees, black raspberries, currant bushes, strawberries, and countless edible herbs and perennial flowers.

Make like-minded friends at a community garden.

I then had the confidence to start a community garden and make new friends. Taking my degrees in biology and education in a different direction than I had originally intended, I started teaching classes and workshops on various aspects of growing food. I also began writing about growing food in residential spaces on my website, TenthAcreFarm.com. Not only had I found a productive way to contribute to my household that I loved, but I was also learning useful skills. Marketable skills. It sparked the entrepreneur in me.

THE SUBURBAN MICRO-FARM

My identity crisis was over! Growing food for my family, preparing it and preserving it, teaching, and writing about it all seemed like something I had done my entire life, and the most natural thing for me to want to do. I finally felt satisfied and useful. In just a few years' time, I had createda micro-farm, which we call Tenth Acre Farm, in the middle of the most typical, American-as-apple-pie suburb.

But here's the catch: Growing food for a household can be a full-time job (the traditional job of the farmer), and so can preparing and preserving it (the traditional job of the farmer's wife). When I added professional aspirations of teaching and writing to my schedule, I instantly felt overwhelmed by all there was to do in the garden and kitchen. What came next was developing a process for making micro-farming possible for busy suburbanites, myself included.

A plum tree fits in a small backyard.

It also occurred to me that after having participated in CSAs (Community Supported Agriculture) for many years as a responsible *consumer*, my new experience as a micro-farmer firmly situated me in the role of *producer*, forcing me to really think about the question, "How do we feed ourselves?" This is a question that garners little attention in today's modern world, mostly because half of all people in the world today live in cities—away from rural food production—where food simply comes from the grocery store.

However, I believe that the suburbs are perfectly positioned to help us feed ourselves into the future as more and more farmland turns into housing subdivisions.

The Suburban Problem

It might seem implausible for the suburbs to be a viable source of food. After all, they are best known for their strip malls, wide expanses of chemically infused lawns, and for consuming far more than they produce. According to the book *Superbia!*, a University of Wisconsin study revealed that there are fewer than 5 acres of earth available to meet the needs of each person on the planet, yet the lifestyle of a typical suburbanite requires over 31 acres of land to accommodate the farms, mines, fields, forests, houses, and roads necessary to serve their voracious appetite for consumption.

This statement about the excess of suburban living makes me feel slightly uneasy. After all, access to cultivable land is the stuff of wars the world over—something to kill for. Yet here we are in the suburbs flaunting our lawns like an expensive hood ornament. The psychologist Erich Fromm says, "The American Dream is another term for crazy." I wondered if I was the only suburbanite who wanted off the crazy train; who perhaps wanted to meet more of their own needs within their current space.

Then I discovered that half of all Americans now live in the suburbs. Bill Mollison said, "The greatest change we need to make is from consumption to production, even if on a small scale, in our own gardens. If only 10% of us do this, there is enough for everyone." Even if just 10% of suburbanites transformed their yards into micro-farms, what a powerful leverage point for change—for both reducing our consumption and becoming producers.

We tend to point fingers at industrial agriculture for their overuse of fresh water, but the fact is that America's largest irrigated crop exists right under our feet in the suburbs. You guessed it: grass. How we use suburban lawns is critical for the health of our local ecosystems.

That probably all sounds well and good, but you may have your doubts about the usefulness of small gardens. After all, before I got started with my edible yard I mistakenly took many myths

about growing food as fact. Here's a look at those myths and some examples of how I busted them—quite delightfully by accident.

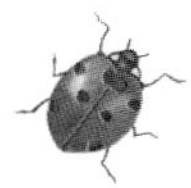 Suburban Micro-Farming Myths

MYTH 1: You need a lot of space to have a productive homestead.

In actuality, it doesn't take much space to grow a lot of food–only creativity. In the next section, you'll see the results of a study that demonstrate how one acre of suburban garden can be three times more productive than one acre of farmland.

My 0.10-acre micro-farm, for example, produces enough homegrown food to incorporate into 50% of our meals. Tip: Plant the parking strip. We planted three dwarf cherry trees in our parking strip (between the sidewalk and the street), which produced 27 pounds of cherries in one year. See chapter 10 for more details about planting the parking strip.

> THE MORE LIMITATION AND RESTRICTIONS YOU PUT ON A DESIGN, THE MORE CREATIVE YOU BECOME.
>
> *GEOFF LAWTON*

MYTH 2: Farming is ugly and should be confined to the backyard.

Growing food can be more imaginative than simply planting a rectangular garden plot in rows. Since our backyard is shady, we grow most of our fruit in the edible front yard. With beautiful leaves and fruit, the strawberries, currants, black raspberries, and cherry trees compete with ornamental landscaping for beauty. In fact, the berry hedge lining our front porch produced 13 pounds of berries in one year and attracts hummingbirds when flowering.

We discovered that edible landscaping can be done the wrong way, and we made plenty of mistakes. See chapter 10 for more edible landscaping ideas and to learn from our mistakes!

MYTH 3: You need flat, sunny land to grow food.

Because most suburbanites didn't choose their property for its farming merits, it is likely that your property has some challenges such as sloping land or shady areas. Have no fear! Land with contours can actually be beneficial for growing food by creating microclimates of hot/dry and cool/moist conditions, allowing you to grow a diversity of crops. See chapter 11 for more ideas on how to take advantage of sloping land.

A partially shady yard can still produce an abundance of food using cool-season vegetables such as leafy greens and root vegetables. In one year, we produced 80 pounds of cool-season vegetables in just two 4x8′ raised beds in our shady backyard. See chapters 4 and 5 for more ideas on working with shade.

Red currants grow in the shady edible landscape.

MYTH 4: Only full-time micro-farmers can yield a lot of produce.

It's true that the more time you spend doing something, the more you get out of it. But that doesn't mean part-time micro-farmers and weekend warriors can't pull off a successful garden. We're all busy, which is why I suggest spending 15 minutes a day in the garden.

Committing to gardening for 15 minutes a day does a lot of things: It keeps gardening as part of the daily routine even when you're busy and inclined to put it off until you have a big block of time (which rarely comes). The practice eases anxiety about what "should" be happening in the garden and connects you to the joy of the growing process. To read more about the 15-minutes-a-day strategy, see chapter 2.

MYTH 5: A farm must have farm animals.

The vision of a traditional farm—with grazing livestock and rows of crops—is a beautiful one, and quite alluring to many micro-farmers who figure out how to raise animals thoughtfully and conscientiously in smaller spaces. However, just as there are traditional farmers who special-

ize in lettuce or corn, for example, micro-farmers shouldn't feel pressured to keep animals unless they really want to. In fact, busy micro-farmers who want to keep livestock should make absolutely sure that they have time in their schedule to properly research and care for the animals.

 The Suburban Micro-Farm Solution

Humans eat a lot of food, which has to come from somewhere. That somewhere is usually far away, industrial farms on wide expanses of clear-cut, rural land. By transitioning annual food production to the suburbs, we open the opportunity for rural lands to focus on things that need more space, such as staple crops, livestock, and timber, and to once more support native flora and fauna species that require large tracts of contiguous, undisturbed land for survival.

In a country where less than 1% of the population farms, increasing food production in suburban landscapes would increase national food security. But I wondered: Would we need more land for food production if we shifted production to the suburbs? Modern industrial agriculture claims to be really efficient at producing food.

The answer is no. According to "The Garden Controversy," a study published by the University of London, suburban backyard gardens are three times more productive than farmland. When comparing food production on one acre of suburban land to one acre of farmland, the suburbs out-produced rural land by three times the amount of food by weight. In other words, small farms and gardens are not a waste of time or space!

Another study also confirms the productive opportunities of the suburbs. An Ohio State University study discovered that the city of Cleveland, Ohio, could produce almost 100% of its needs for fruits, vegetables, chicken eggs, and honey by using vacant lots, commercial and industrial rooftops, and just 9% of each residential property for growing gardens, raising chickens, or keeping bees.

At first it may seem unattainable for any city to achieve the amount of self-sufficiency referenced in this study, but I am hopeful that, if a land-scarce city could achieve productivity levels rivaling rural operations, the suburbs—with our wide expanses of lawn—could certainly scale up food production and "take more responsibility for our own existence," as Mollison urges us to do.

We've already experimented with the small-scale model of food production in the form of Victory Gardens. During the World Wars I and II, many farmers left their fields to become soldiers, radically suspending food production. Citizens were encouraged to support the war effort by

growing food in Victory Gardens and preserving it for winter. Plots were created anywhere land was already cleared: farms, backyards, private estates, company grounds, city parks, vacant lots, schoolyards, and even window boxes and rooftop gardens. At their peak in 1944, 20 million garden plots produced 40% of all produce consumed in the United States.

As I mentioned earlier, we're able to produce enough fruits, vegetables, and herbs to incorporate homegrown food into 50% of our meals on our tiny, 0.10-acre lot.

Suburban micro-farmers could produce at least some food to feed their own families and reduce their reliance on food shipped long distances to the grocery store. More and more, there is concern about the growing practices in those faraway industrial farms, as well as whether those in charge have our best interests in mind. When you grow your own food, on the other hand, you know what you are getting. Additionally, it's clear that transitioning the household from a unit of consumption to one of production can reduce costs.

Whether improving your family's food security, safeguarding their health, or saving money is your interest, growing your own food is a worthwhile endeavor.

Many of the original suburban houses developed in the 1950s came with three fruit trees. What an amazing gift to have so much delicious abundance growing right in the spaces where people live! Sadly, this practice hasn't continued into modern housing developments. Today, our suburban landscapes are more sterile than ever. In a transient culture where one-sixth of the population moves each year, we could leave our properties better than we found them by simply planting a fruit or nut tree for the next occupants—and the next generation.

Transitioning a large part of food production away from rural areas to suburban lawns carries a lot of potential, but great care must be taken to ensure that micro-farming practices regenerate fertility and ecological diversity rather than further degrade soil quality.

Toby Hemenway, author of *Gaia's Garden*, writes that we must use "...time-tested techniques honed to perfection by indigenous people, restoration ecologists, organic farmers, and cutting-edge landscape designers." A permaculture approach to suburban micro-farming will help reduce the

toxic use of chemicals, improve soil fertility, reduce erosion, increase habitat for pollinators and beneficial insects, create wild areas, and increase food productivity.

There is a permaculture principle that states, "The Problem Is the Solution." This principle encourages us to consider that when we think of something as a problem, it may be so only because of the way we think about it. For example, some people today think of the suburbs as an embarrassment, with water-hoarding lawns and a lack of car-centric alternatives, so they choose to live elsewhere. I once made that decision for myself and enjoyed my urban apartment within walking distance of amenities and no lawn to worry about. It was a dandy time!

However, if we realize the enormous potential the suburbs have to change overall consumption habits and transform land use practices, the suburbs could end up being just the solution our cities—and perhaps even civilization—need. After all, out-lying villages have performed this function in ancient cities of the world throughout history. All we need are some pioneering micro-farmers!

As the population of the country swells to unprecedented numbers and once-fertile land is gobbled up and turned into housing developments, it is clear that we need a new kind of farming to match the realities of modern life. The following pages will help you pioneer your own suburban micro-farm, so you can turn problems into solutions!

2 MANAGING EXPECTATIONS

BLOOM WHERE YOU ARE PLANTED.

AUTHOR UNKNOWN

When you start dreaming of a new micro-farm, or dream of expanding an existing one, one of the most common mistakes is thinking too big. This is one of those cases where the phrase "Go big or go home" does not apply. To avoid getting overwhelmed or disillusioned, it will be imperative to manage expectations wisely, as well as to properly identify timelines and what lies ahead to minimize surprises.

Another pitfall to avoid, especially in the suburbs, is the tricky subject of dealing with neighbors and their own expectations. You need to manage those relationships well, especially when the neighbors have different expectations for your lawn and might not understand your farming aspirations. Along those lines, many of us will seek out additional growing spaces when our own yards don't provide enough. That's where community gardens come in.

In this chapter, we'll look at how to prevent *Garden Overwhelm Syndrome*, master some life hacks and tips that will help you manage your garden on a busy schedule, learn how to deal with failure, and know what to do if you've tried the life hacks and still can't manage the garden. If you have doubts about how you'll manage it all, don't bet the micro-farm yet. The information in this chapter will help to set you up for success.

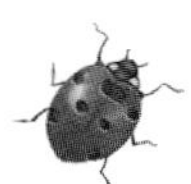

Prevent Garden Overwhelm Syndrome

We've all done it: From our cozy winter table, it's easy to plan an elaborate garden that looks completely manageable on paper. Planting season gets underway and then *bam*! The weeds of late spring hit, and we're goners. *Garden Overwhelm Syndrome* takes over, and we throw in the towel. They say prevention is the best medicine, so here are my tips for preventing overwhelm from ruining your garden season.

Get it on the calendar.

My **monthly calendar printouts** [get them at www.TenthAcreFarm.com/tsmf-companion] allow you to schedule your entire gardening year, much like your family's master activities calendar. Once filled out, you'll be able to see what each month has in store for you, from when to start tomato seeds inside, to when to sow carrot seeds outside, to when to plant garlic or fruit trees. Based on what you want to grow, discover what your time commitment will look like before the garden season even begins (reality check!). The printouts will aid in making adjustments so that you can be successful. See chapter 7 for more details about using the monthly calendars.

TIP: MONTHLY CHECKLISTS

The **monthly checklists** (look for them at www.TenthAcreFarm.com/tsmf-companion) will help you prioritize and make efficient use of the time you have to spend in the garden. For more information about using the monthly checklists, see chapter 7.

February						
Sunday	Monday	Tuesday	Wednesday	Thursday	Friday	Saturday
	1	2 Harvest beets, carrots, kale & spinach from cold frame.	3	4	5	6 Tidy garden - cut back dead plants, weed, rake leaves.
7	8	9 Start broccoli and kale indoors.	10	11	12	13 Build a compost bin.
14	15	16 Start peppers indoors.	17	18	19	20 Add soil ammendments to spring cold frame garden beds.
21	22	23	24	25	26	27

Plant what you love.

If you're new to gardening or new to managing an abundance of fresh produce in the kitchen, I strongly recommend growing only what you know your family loves. While you're adjusting to making time in your schedule for gardening and the extra kitchen work, it's better to not worry about veggies that will seem like a chore to prepare and eat.

Prioritize harvesting.

Capitalize on your hard work—don't let the produce you've worked so hard to grow all season go to waste because you're planting for the next season, catching up on weeding, or otherwise letting good food rot on the vine. Always harvest first!

Become a harvest manager.

Harvest on time (it bears repeating!) and make sure that the harvest gets used in the kitchen. Experiment and develop a kitchen routine for using and preserving the existing harvest. You'll be better equipped to handle that large garden with large harvests in the future if you first take the time to master a food prep and food preservation routine now.

Maintain the existing garden.

Plant only what you can manage – if you can't maintain the existing stuff, why plant more? Gardens can be overwhelming, and that's why I prioritize keeping what I've already planted alive and harvesting what I've already planted before I take on new tasks like building more garden beds or planting for the next season.

Work toward the right time.

It's easy to look into the future and see a giant productive garden. When you dream about this future giant garden but have an unrealistic view of how much time you have in your daily life to accomplish it, your priorities can easily get a little skewed. (I know this from experience.)

If you don't have more *time* available to you, a bigger garden may not be the best option. Instead, take the opportunity to become a more efficient and organized micro-farmer. Focus on fine-tuning your routine and skills that will help you master the bigger garden when you're ready. Commit to a regular planting and maintenance schedule.

Life Hacks for the Busy Micro-Farmer

I'm not gonna lie: This farming thing is a commitment. Many of us genuinely want to have our own micro-farm and to supply our family with fresh, nutritious food. The only problem is that real-life stuff always seems to get in the way.

Do you know someone who seems to do it all? The SuperMom who works full time, coaches little league, bakes cookies to welcome the new neighbors, and still manages to look good? Perhaps it is a little deceiving, for no one can truly do it all. Something eventually gets missed or left undone.

However, I also believe that those who reach seemingly impossible goals in these crazy, modern times are those who have discovered life hacks that streamline their journey to success and accomplishment.

Life hacks do four things:

1. They help focus your attention on what you really want.
2. They improve your chances of establishing a routine to achieve success.
3. They reduce the amount of brainpower needed to accomplish a goal.
4. They improve efficiency so that the time required to accomplish a goal is reduced.

In short, life hacks emphasize success over stress. They also leave little room for excuses.

Life Hack #1: 15 Minutes a Day

I like to get outside a minimum of 15 minutes each day. Whereas some people might go for a daily walk or jog, the garden is often both my daily exercise and meditation. A daily visit allows me to be in tune with the garden, giving me the best chance to notice pest or disease problems at their onset as well as any other subtle changes.

> THE RHYTHM OF MY DAY BEGINS WITH A CUP OF COFFEE AND A LITTLE BIT OF WEEDING OR DREAMING.
>
> *BETSY CAÑAS GARMON*

The more time you spend there, the more you have the opportunity to connect to your land and get to know its nooks and crannies. This is why one acre of suburbia has been shown to be more productive than one acre of farmland. You can really get to know your small piece of land when you're connecting to it every day, season after season.

Now, to be certain, the more time you commit to your micro-farm, the more you get out of it. But don't let perfect be the enemy of good. It's important to be a little Zen about this suburban micro-farming thing and only do what you can do. Just 15 minutes is a great place to start.

Are you so busy that you don't even have 15 minutes to spare? I've been there! When I was working outside of the home, I had to be creative about how I squeezed an extra 15 minutes out of my day for garden time. So I came up with *Life Hack* #2.

Life Hack #2: Split Personality

If you work away from home each day, consider dividing the 15 minutes into two pieces (7 minutes + 8 minutes):

The 7-Minute Morning: Grab your coffee and set a timer for seven minutes in the morning for watering, weeding, or doing anything else on the monthly checklist. You may even have time to throw some seeds in a new bed!

The 8-Minute Evening: Grab your happy hour drink of choice and spend eight minutes in the evening working through the monthly checklist. You might even get to harvest something for dinner!

Harvest strawberries with your morning coffee.

Life Hack #3: Weekend Micro-Farming

It's possible that the 15 minutes-a-day plan just isn't the ticket for you while you juggle work, kids' schedules, etc. No problem! You'll have to block off a chunk of time each weekend and do it all at once. An hour and 45 minutes each weekend should get you started. Bonus points for extra time!

Life Hack #4: Go Small

Plant a smaller garden, such as one 3x6 raised bed. It will take more time as a beginner to learn the ropes, so keep it manageable. When you're confident that you can manage one small bed, add another. As Mother Teresa said, "Not all of us can do great things. But we can do small things with great love."

Life Hack #5: Avoid Seedy Affairs

Avoid planting seeds in the garden. Directly sown seeds need extra care to ensure germination, which you might not have time for. Beans, peas, and root vegetables are examples of crops that are directly sown. If you do sow them anyway, be prepared to find joy in whatever comes up rather than disappointment in what doesn't.

Life Hack #6: Farm Out the Work

Instead of managing an expensive and time-consuming indoor seed-starting operation, opt to buy seedlings from your local nursery or farmers' market instead.

Life Hack #7: Perennial Beauty

Grow perennials. Perennials such as asparagus, rhubarb, fruit trees, nut trees, berry and hazelnut bushes, strawberries, and many herbs all come back year after year without much work on your end.

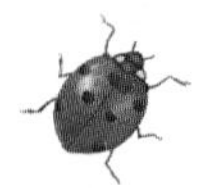

When It's Not Enough

Let's imagine you had visions of growing all of your household's produce, cooking homegrown meals all season long, and preserving the rest for winter eating. You worked hard all spring to try to stick to a schedule of planning the garden, preparing the soil, and starting seeds, but somewhere along the way you fell off the wagon.

You're realizing that this *grow your own food* thing takes a lot more commitment than you can really give it right now. Above, I offered some suggestions for hacking the micro-farmer's life on a busy schedule, but it turns out your life necessitates an even more basic set of *get started* recommendations. Your top priority needs to be—first and foremost—eating fresh, chemical-free produce and keeping your busy family healthy. In that spirit, here are my top recommendations for busier-than-busy micro-farmers. They also happen to be excellent options for those of us without any yard space at all.

Join a CSA

CSA stands for Community Supported Agriculture, which is a way for consumers to buy farm-fresh food directly from a local farmer. While CSAs aren't available in all areas of the country, the probability that a CSA is available in suburban areas is high. You might need to do some Googling or check your local farmers' market, but the chances are good you'll find a farmer who offers a program.

CSAs come in all shapes and sizes. Eight years ago, when Mr. Weekend Warrior and I decided we wanted to grow all of our own produce, we didn't start by creating a garden. Rather, we started in the kitchen by joining a CSA.

CSA PRODUCE IS FOR YOU

CSAs are a boon for experienced gardeners who can't seem to find time to keep a garden. A CSA will keep your family eating healthy produce and will help you establish relationships with like-minded foodies in your local area.

Let me explain why this was the most amazing first step we could have ever taken. Through the CSA, we received more produce from the local farm during that first year of membership than we have ever seen before. We didn't even know what half the produce items were. (*What is kale?* Don't judge!) It took most of our free time to research how to use everything, prepare the food, and preserve the excess, which is excellent training for when you finally pull in your own big harvest.

I felt relieved that I wasn't trying to figure all that out while also figuring out the gardening side of things! The coolest part was that the CSA we joined had a "work" component, where we worked a certain number of hours on the farm with the farmer or other CSA members as part of our payment. It gave us a chance to learn gardening skills and meet loads of other people in our own community with the same interests.

By choosing to do only one thing (kitchen prep), we still got loads of healthy veggies in our diet without trying to fit in time to grow it all. As we constructed more and more garden beds in our own yard (starting small and increasing the garden space over time), the lessons from the CSA—both in the kitchen and in the farm fields—were priceless.

Join a Community Garden

Community gardens—at least in the United States—have the stigma of being used only by disadvantaged urban residents. Although this is certainly true in many areas, I fought hard during my time as founder and coordinator of my local suburban community garden to dispel this stigma. I believe community gardens can be of great benefit to all communities. They are not only assets to landless urban folks—rich or poor—but they can also benefit those in suburban areas who are challenged by shade, deer, and other challenges in their own yards, and who feel isolated in their car-centric neighborhoods.

Friends at the community garden learn together.

But let's get to the root of why community gardens are useful for super-busy micro-farmers. The thing I love about a community garden is the fact that the amount of space is defined. A small 3x6 or 4x8 raised bed is easy to maintain when it's not in your backyard, and here's why: The raised bed in my backyard is *in my yard*, which means every time I go to take care of it I see all the other landscaping that needs maintained, that I don't have time for. I feel overwhelmed, and it makes me want to abandon all of it.

The community garden doesn't feel like work, however. It's like a mini-escape from the craziness of life! Visit your plot an hour a week, and while you're there, you might have company from fellow gardeners. It's so much fun to pull weeds and shoot the breeze at the same time, with the unkempt backyard out of sight, out of mind. For this option to be really successful, pick the same hour every week and write it on your calendar so that it's a weekly, non-negotiable hour to yourself.

> IT TAKES A WHILE TO GRASP THAT A GARDEN ISN'T A TESTING GROUND FOR CHARACTER AND TO STOP ASKING, WHAT DID I DO WRONG? MAYBE NOTHING.
>
> *ELEANOR PERÉNYI, GREEN THOUGHTS, 1981*

I hadn't realized how much I enjoyed the social aspect of gardening until I started a community garden. Gardeners at my community garden occasionally celebrate our bond with Wine Down Wednesday—a new meaning to happy hour gardening! The camaraderie truly does make the work seem so much lighter and go so much faster, and it's a great way to meet like-minded friends. I will talk more about my community garden experience later in this chapter.

Whether you join a CSA or a community garden, know that there are always options for you to continue to work toward your goal of providing your family with fresh, healthy produce while learning productive skills at the same time.

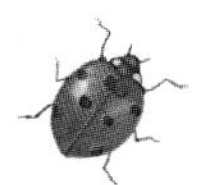 Overcoming Brown Thumbs

The term *green thumb* is a little misleading. It implies that you are either born with the innate knowledge of growing things, or you aren't. On the contrary, I think we're all born with the ability to learn

any craft—including how to grow things well—if we're willing to spend time practicing. In the book *Outliers: The Story of Success*, author Malcolm Gladwell tells about the "10,000-Hour Rule": The key to becoming an expert at something is simply a matter of practicing a task for a total of 10,000 hours, which is about 20 hours a week for 10 years. Practice makes perfect, as they say.

When you're first starting out with a new craft, it's easy to worry about whether you're doing things the "right" way, when really you just need to have fun and learn by experimenting. Gardening is no different. What if you put a "curious kid" hat on and became a detective? What if everything in the garden was a mystery that needed solved?

When something goes wrong, such as your sweet potato vines not growing tubers, get out a little notepad like Penny from *Inspector Gadget* (have I dated myself?) and take notes like this:

- Sweet potato leaves are yellowing and stunted.
- Soil is compacted and heavy.
- Location is shady.
- Soil near the garden is disturbed. (*a child's footprint or signs of a groundhog or vole that loves sweet potatoes?*)

Even though it might seem silly (shouldn't life be silly anyway?), this detective process can help you get to the bottom of why your sweet potatoes weren't growing any tubers. Do some research on each symptom, what might cause it, and what to try next time to improve. Rather than failing, or not having a green thumb, simply solve the mystery. Like learning to ride a bike, failure is an opportunity to learn and practice. Hitting that 10,000-hour mark is going to take a lot of practice!

When I first began learning about vegetable gardening, I constructed two 10-x-10-foot gardens on a friend's property as part of a landshare project. I sheet-mulched the areas (cardboard covered with imported compost soil), fenced them in from deer, and for the first time in my life I planted basil, beans, carrots, corn, cucumbers, peppers, potatoes, radishes, sweet potatoes, and tomatoes. I was so excited to have my own little gardens to experiment with and grow whatever I wanted! A labor of love, I weeded and watered them regularly, then anxiously awaited a harvest.

Everything started producing, and then on July 30th, I noticed one armyworm and one cucumber beetle. By September, pests had decimated both gardens.

I had done everything by the book, so why was I getting these pests? Was it my brown thumb? Through detective work, I discovered that this was common in new gardens because the soil life—the foundation of a healthy garden—was getting acquainted to the new conditions. My visions of healthy crops and abundant harvests would have to wait while the beneficial soil organisms were settling down in their new home. No pesticides—natural or chemical—would have saved my harvest; giving my gardens a year to settle was the only lasting solution.

DO SOMETHING,
EVEN IF IT'S WRONG.

JOEL SALATIN

Through failure, I learned a valuable lesson about soil biology, which I have referred back to many times when establishing new gardens. Brown thumbs are an opportunity for learning, and failure is not an indication of your gardening ability. Don't be afraid to make mistakes. This is a learning process. If something doesn't go as planned, simply enjoy the experience, make some notes about what happened and what you might try next time, and move on.

Remember, life is about the journey. It should be full of silliness and fun as you practice your way to green thumb expertise!

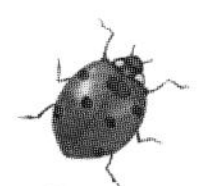

Setting Up Your Neighborhood for Success

Micro-farming in the suburbs is a grand social experiment. While farming is a tradition and mainstay of the wide-open spaces of rural areas, it is certainly not the tradition of lawn-centric suburbs, where differing expectations may collide at the property line. As you go down the path of pioneering a micro-farm in the suburbs, you will have to manage neighbor relations very carefully.

While I hope all of your neighbors enjoy your beautiful, productive landscape, it might not be all roses. The suburbs have a culture all their own, and in permaculture we talk about this culture as an "invisible structure," an intangible—yet essential—ingredient that affects our motivation to start and maintain a certain project.

Have you seen the movie *The 'Burbs* starring Tom Hanks? It's a spoof on suburban living, in which Tom Hanks' character and several other neighbors begin spying on new neighbors because it appears they might be up to something weird. Although the new neighbors in the film were legitimately spooky, the social politics of standing out in a neighborhood can be disastrous.

In 2010, we began transforming our front yard into a productive landscape, which meant being outside and interacting with neighbors more. They would gather in our yard and chat while we worked—something that had never happened prior to this point. A year later, as we dug a giant hole in the front yard to create a rain garden, the neighborhood kids wanted to know what we were up to (What young boy can resist the excitement of digging a big hole?). We told them we were putting in a swimming pool, and they stopped by with amusement many days after school to watch the transformation of the front yard "pool."

Not only did we dig up the yard, but we also didn't treat our lawn. The strip of grass that remained in the front yard after we built the gardens was carpeted with dandelions that spring. One day, an older gentleman down the street—who keeps an immaculate lawn—stopped in his car. Rolling down the window, he said, "My lawn care guy is stopping by today to spray my weeds. You want me to send him down when he's done?" I replied that it was a generous offer, but that I really enjoyed the beauty of the dandelions. He muttered something under his breath and drove off.

I installed a vegetable garden in my neighbor's backyard and share the harvest with him and other neighbors.

I continued to wave and be cheerful whenever he passed by our house, and once summer rolled around, the big holes in our front yard were now leafed out with beautiful flowers and vegetables. This time when he stopped his car, he did so to compliment the beauty and to marvel at the vigor of the tomato plants. That was certainly a lesson in patience, as I had felt very frustrated by his attitude earlier in the season.

Being in the front yard more often gave me more opportunities to chat with neighbors,

something I felt was important, but I am no natural conversationalist. Luckily, I didn't have to do all the work. Another couple came down to introduce themselves, bringing a bag full of their home-grown produce to trade for our gorgeous front yard heirloom tomatoes. I learned that we weren't the only ones who thought the front yard garden was cool.

One neighbor with a particularly immaculate lawn became decidedly less friendly when we dug up the front yard. It seems public opinion carries a lot of weight with him, however, because a year later we hosted a class of 25 people who came out to take a tour of our edible front yard. Afterward, the neighbor approached me and wanted to know more about our style of landscaping, evidently convinced that our methods were acceptable now that there was social proof. Just like in *The 'Burbs,* social proof matters. If a group of people had wanted to learn from us, then we must not be too crazy after all! He returned to his former friendly self.

A middle schooler who started his own lawn business, as well as a couple of other neighbors, started bringing their grass clippings and leaf bags over to our compost.

Still another backyard neighbor—who was originally leery of our anti-lawn activities—in the end volunteered a portion of his yard for me to install a raised bed, for which I enlisted the help of a few neighbor kids.

These are just a few of the experiences I've had in engaging with neighbors about our desire to grow more edibles than grass. Neighbor relations are not always positive experiences, though. Here are some tips for bettering relations with neighbors and to keep from burning bridges.

Neighbor Relations

When I started learning about all of this micro-farming stuff, I was excited to dig up the lawn and create a productive homestead. But I was also a little terrified of what the neighbors would think. This fear turned into irritation about the fact that none of them seemed to have an inkling about this *farm your yard* movement. I found myself wishing they would get on board, or that I had different neighbors so that I didn't have to always explain myself or feel so different.

When I engaged with my neighbors, I wanted them to delight in what I was doing, take interest, and maybe even go home and try it themselves. I wanted to explain why I was doing all of this. I envisioned a new culture of social interaction on my street, one that included meaningful activities and conversation.

But that wasn't happening in the way I had imagined.

In *The Charisma Myth: How Anyone Can Master the Art and Science of Personal Magnetism*, author Olivia Fox Cabane tells the story of the 1886 prime minister race in the United Kingdom. A young woman was lucky enough to have dinner with each of the two prospective prime ministers in the last week before the election. Afterward, she was asked about her impressions of the two men.

She replied that the first candidate was "the cleverest person in England," but the second candidate made her feel as though she herself "was the cleverest person in England."

Cabane goes on to say that when we're charismatic, we make other people feel a certain way—"intelligent, impressive, and fascinating." In turn, they will want to be a part of our world—to learn about our perspective, to give it a try, to participate and support us.

In the case of my neighbors, I realized that the problem was not theirs, but an attitude problem on my end. I wanted them to take an interest in me, but I hadn't taken an interest in them. I hadn't spent any time talking to them about whatever topics they were enthusiastic about. I hadn't made them feel "intelligent, impressive, and fascinating."

I finally learned that the only way to coexist (and stop having unreasonable expectations that they were going to join my farming crusade) was to be happy and cheerful, share my harvest, and share in their lives as human beings, without the slightest care that we have absolutely nothing in common. That's when the neighbors began to express curiosity about our farming actions regardless of our differences in lawn management strategies.

Cherry trees bloom in the parking strip.

That's not to say that things are all hunky-dory. 2011 was the year we planted three dwarf cherry trees in the parking strip—that unused strip of grass between the sidewalk and the street. One neighbor cautioned us: "You put that in the front yard, the neighborhood kids are going to steal the fruit!" To which I replied, "Great!" I imagined how lovely it would be to see the kids pulling cherries out of the trees rather than chasing down the ice cream truck.

That conversation motivated me to go one step further: tell the neighbors about the fruit trees and encourage them to harvest if they saw fruit. After all, although homeowners are responsible for maintaining the parking strip, it is also public property.

The problem is that fruit trees don't produce fruit right away. In the first four years, we only got a few handfuls of cherries, which wasn't very interesting to those who are accustomed to buying cherries whenever they darn well please at the grocery store. Then, in the fifth year, the trees came into their own and we hit the mother lode. The trees were dripping with fruit, and we harvested 27 pounds of cherries.

I was so ready to see kids on bikes and dog walkers alike joyously helping themselves to handfuls of the sweet-tart goodness. But you know what really happened? By year five, neighbors weren't so keen about the open invitation to harvest from our yard. We missed the window of time where they were open to the spirit of sharing.

Now that there were cherries available, many were afraid of them (*Are they poisonous?*), one asked if they were apples, and a few said they didn't like cherries.

The moral is that if a community ambiance of sharing and appreciation of homegrown food is what you're after, it might work and it might not. There is no magic bullet to guarantee excitement. Cultural and social norms are neighborhood specific. The best you can do is to continue to model cheerful enthusiasm and feel grateful that you are adopting a lifestyle that is meaningful and fulfilling for you. Like Van Gogh, who only sold one painting his entire life, you may have to settle with the knowledge that others may not immediately appreciate your efforts.

While I hope all of your neighbor relationships are the stuff of fairy tales, you may have to deal with a few pesky types that enjoy spraying their side of the fence with herbicides and pesticides. After all, many of them attended the church of lawn worship.

Who can blame them? Most of us grew up going to the same church. All is not lost, however. Here's a look at how we interacted with our neighbors and what you can do if yours aren't impressed with your chemical-free farming interests.

In 2010, when we began transforming our front yard into a productive landscape, we started interacting with neighbors more, and it seemed like we would live happily ever after. Since we were getting along so well, Mr. Weekend Warrior asked our neighbor—whose yard drains into

ours—if he might be interested in stopping the chemical lawn treatments, or switching to the organic version. We offered to pay the difference in price. Mr. WW explained to the neighbor that we were growing edibles, and we were trying to do it without chemicals.

The neighbor replied, "You know, I've been getting the lawn treatment for years out of habit. I don't even know why I was doing it. I'm just going to cancel it." It was the most amazing response we could have hoped for. I gave him some tomatoes (his favorite homegrown vegetable) that year as a thank you.

Generally, people want to know how something they are doing affects you. Stick to the facts: "We're growing edibles. We're trying to do it without chemicals, but your yard drains into ours, so is there anything we can do about this? Here is what I'm willing to do to meet you in the middle." Don't use guilt. If they were the type of people to be motivated by environmental degradation or human health risks, you wouldn't need to have a conversation with them about chemicals to begin with!

Chastising neighbors for their actions—which, of course, affects others downstream or downwind—is unlikely to make them feel like helping you out. If chatting them up with jovial—but direct—commentary doesn't encourage your neighbors to stop applying chemicals, then you might need to accept your differences and set up some physical barriers to protect your growing area.

Mounded garden beds along property lines will act as a barrier and keep fertilizer from draining into your gardens. The mounded bed will draw up and filter contaminated water. Plant the mound with a (non-edible) hedgerow for even more filtration, and to act as a barrier from aerial spray drift. Swales are a technique for catching and filtering rainwater, and they create the perfect mounded garden bed on which to plant hedgerows. (For more details about hedgerows and swales, see chapter 11.)

Children tasting fresh mint at the community garden

Don't forget to smile and share your harvest!

In order to maintain neighborly relationships, do what makes you happy on your own property, but be cognizant of how it may affect your neighbors. Try not to judge

others. Until laws require homeowners or the producers of lawn care products to take more responsibility for their polluting actions, you'll need to compromise, negotiate, and model what you wish to see in the world while allowing them to continue the absurd act of using toxic products.

Don't let me fool you—I'm no expert at any of this. Neighbor relations can be uncomfortable, but learning to negotiate and compromise are valuable skills that are useful for any aspect of life.

Adventures in Community Gardening

Earlier in this chapter, I mentioned community gardens as a solution for those with limited time and space. While they are common in urban areas, they have only recently become more commonplace in the suburbs. No matter your location, a community garden can be a great way to meet like-minded neighbors, learn and practice productive skills, and strengthen community bonds.

That's exactly what I had hoped to accomplish in 2011, when I decided to bring the concept of a productive suburban micro-farm to my community. In a township without a town center or any walkable, public spaces, I envisioned a community garden as both a gathering space and a demonstration site of ecologically friendly food growing methods.

The first challenge was to find a space. After many disappointments, I landed an eroded hillside behind the local university. The site was surrounded by woods and completely hidden from view of anyone who would look for it. To say that it wasn't ideal was an understatement.

The community garden model of a flat, sunny area scattered with neat rows of raised beds quickly shifted into a vision of a community hillside food forest in which permaculture design fused together ecological restoration and food production. Rather than rent out raised beds, our community gardened the space cooperatively as we built terraces together and learned to stabilize the hillside while also getting an edible yield.

We set up regular communal gardening days for Wednesday evenings and Saturday mornings, and shared the harvest with whoever showed up. We made participation free so there weren't any financial barriers. This setup was meant to benefit the typical suburban family who can't commit to one more activity in their busy schedule, but would like the leeway to be able to show up when they can.

I love this model in theory, because in a suburban community where most residents have their own yards, the emphasis can be less on high production, and more on working together to learn useful skills—a demonstration, or experiential garden.

Hillside Community Garden and Food Forest

A community leek harvest in December

Five years and 4,000 hours of work later, the hillside—once filled with garbage and overgrown with invasive honeysuckle—is a sight to behold with 15 terraced garden beds growing fruits, vegetables, herbs, and flowers. A compost center, sitting area, ornamental welcome garden, and shed complete the space. Personally, I made several lasting friendships and learned a boatload of skills that have made me a more confident gardener, educator, and permaculture practitioner.

Theories don't always match reality, however. While I accomplished two of my three goals (meeting like-minded neighbors and learning productive skills), we completely failed at strengthening community bonds. Though we offered classes, workshops, tours, community events, and service learning days in conjunction with the university, we finished our fifth year with only four regular gardeners.

This highlights a downfall of many community projects—strong in theory but weak in reality. Still, it is important to model what we wish to see in the world. Bringing a visual to life is the first step in pioneering an idea, even if we won't live to see the effects. In fact, in the fifth year of the project, township officials approached us for advice and collaboration in starting a township-run garden.

If you are inspired to start a community garden, here are some of the lessons I learned from my experience in founding and coordinating a community garden/food forest project.

Establish three to five co-organizers to help found and initiate the project.

No matter the project, it will be a lot of work to make it successful. Be sure there is a governing body that oversees the project rather than a single individual. That way, if someone needs to take a vacation, tend to an emergency, or leave the project altogether, the project doesn't have to die or go unattended.

In addition to being generally supportive, an organizing team sends out a bigger vibe of energy into the community that will attract more participants, making the project more resilient and more likely to last. Although I quickly formed a support team after establishing my community garden, it didn't have the same effect as having a team that founded it together.

Establish a buy-in from at least two community partners.

If participation is what you're after, potential participants will feel better about contributing to a project that is connected to someone or something they already know. Between yourself and your five organizers, you'll be able to spread the word around your community to a lot of people. A community organization or partnership can attract an even more diverse crowd. For example,

having support from our local government and university were key in promoting our classes and events. Non-profits make excellent community partners.

Encourage deep participation.

Think deeply about the kind of participation your project requires. Between the local university, the nunnery, and the local schools, we had many individuals and groups who wanted to volunteer a single day of service at our community garden. It is enticing, because these types of service events allowed us to promote the partnership in the local press and social media, and the participants get to feel good about a day of service.

Unfortunately, this shallow form of participation doesn't help anyone. In reality, we would spend the entire day of service training the new volunteers how to do the skill at hand rather than completing a project, and they would leave not fully understanding or appreciating the depth of our restoration work. After a while, we started focusing more on creating community with people who were truly committed to the project on a regular basis, and less on the media blitz of working with one-off volunteers.

Be flexible, but find your identity.

Our project started out trying to be everything that the community would expect from a community garden. We wanted to be a center of gardening education, so we promoted classes and workshops. We wanted to be accessible to everyone, so we made participation free. We wanted to be socially responsible, so we created a relationship with the local food pantry in order to donate excess produce. We wanted to promote empowerment in the home garden, so we started hundreds of vegetable and herb seedlings each spring and sold them at our annual plant sale. We wanted to demonstrate to the university that we could be an asset to the academic community, so we committed to speaking to classes, attending events on campus, and accepting service groups on demand. We wanted to demonstrate ecological restoration with an edible twist, so we sought council from local permaculturists.

Ultimately, our small group of part-time volunteers couldn't possibly meet all of these needs of the community. We grew weary of doing the sizeable work of a non-profit organization. Eventually we settled into the basic necessity of our site: ecological restoration using permaculture strategies.

Require a buy-in for participants.

It is said that, in general, people will appreciate something they have paid for over something that was gifted to them for free. In an effort to make our community garden accessible, we didn't

charge a fee for participation. Without that commitment and desire to "get their money's worth," however, community members didn't have the incentive to show up, even though they may have fully supported the project and appreciated it being there.

For most of us, the road to managing a micro-farm will take a winding path as we balance learning productive skills while meeting our real-life needs and financial responsibilities. With the tips in this chapter, you will be able to manage failed expectations, bumps in the road, and neighbor/community relations. And you thought all you needed to do was plant a garden!

DEVELOPING HEALTHY SOIL

TO BE A SUCCESSFUL FARMER, ONE MUST FIRST KNOW THE NATURE OF THE SOIL.

XENOPHON, OECONOMICUS

When I first began my adventure of farming my suburban yard, there was a lot of work to do to regenerate the fertility that had been lost due to years of neglect. Like most residential landscapes, my yard had a story. It had been managed since the 1950s with a goal much different than my modern micro-farming goals. In the heyday of suburban development of the 1950s, lawn was an ornament to signify a departure from the rural, agrarian life. It announced to the world that the family who lived there was well-to-do enough to buy their food at the grocery store rather than grow their own. Lawn provided social status—and it still does.

My front yard lawn, likely pumped with chemicals to achieve the most tidy and sterile appearance, was the pride and joy of the original owners of my home. When it was time for me to carry on the tradition, the soil under this status ornament had been reduced to lifeless, hardpan dirt.

My backyard was once an in-ground swimming pool, and the former owners had simply filled it in with lifeless fill dirt. Today, hardpan clay mingles with large and unmovable chunks of blue-painted concrete under the soil surface. As with all residential efforts to regenerate fertility and return biodiversity to the lawn desert, I had my work cut out for me.

In this chapter, we'll look at how to regenerate healthy soil naturally, how to maintain soil health in the long run, and how to protect this valuable asset.

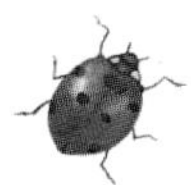

Building Soil Health and Fertility

The lack of healthy soil is a common reason for garden failure. We are often concerned with the outward appearance of our plants as an indicator of health, such as healthy leaves or fruit, yet these are merely symptoms of what's going on below ground. It is the health of the soil that will—more often than not—dictate your plants' chances of success.

I began in my yard by focusing on building humus and bringing life back to the soil. Healthy soil is high in humus—the organic component of soil necessary for plant growth. As farmers, we want to strive to actively build and protect it. Healthy soil will also contain a high number of beneficial organisms that will naturally aerate the soil and develop mutual relationships with plant roots, bringing them nutrients and water, and protecting them against pests and disease.

DETERMINING SOIL TILTH:

Soil tilth is the degree to which soil is suitable for sowing seeds and cultivating crops. Healthy, fertile soil with good tilth has a loose, crumbly structure that allows delicate roots to grow healthy and strong. Spending the time to create healthy soil will pay dividends come harvest time in the garden.

Most of us start out with soil that needs to be improved, and by taking the following steps we can create a cyclic nature of fertility generation.

- **Aerate:** The easiest way to start building fertility in an existing garden is by aerating—not turning—the soil with a digging fork. Aerated soil will absorb more water and attract earthworms and other soil organisms.
- **Add Organic Matter:** Improve topsoil by adding organic matter such as compost soil, worm castings, leaf mold, and aged manure. Add lots of it, every year, year after year. Earthworms and soil organisms will turn the organic matter into humus (the building block of soil), and help the soil to better absorb rainwater and nutrients. When soil is aerated and the quantity of humus increased, plants will be healthier, better able to thrive, and will be more adaptable to extreme temperatures and conditions.
- **Leave Plant Matter:** As plants naturally die back, their decaying roots attract worms and other beneficial soil organisms. Instead of pulling spent plants at the end of the season, cut them off at the base, leaving the roots to decompose. This will reduce disruption of the soil, feed soil organisms, and improve fertility.

  The following year, as you begin to plant new seedlings, you may run into an old root crown from the previous year. Rather than pulling it out, simply plant the new seedling directly next to it. In this way, the old root will actually feed your new seedling.

  If the spent plant was diseased or suffered from a pest, it should be pulled and disposed of completely in the trash or burned, rather than left in place.
- **Encourage Beneficial Fungi:** Beneficial fungi are usually the last to show up at the party. Their existence demonstrates that some measure of self-regulating soil health has already been achieved, and still, they work to improve the soil even more. Beneficial fungi attach themselves to plant roots and extend up to 900 times farther than the plant roots themselves, aerating and actively searching for nutrients and water. They make plants even more resilient against growing challenges. Beneficial fungi encourage more soil aeration, and the cycle begins again.
- **Don't Walk on the Beds:** It is extremely important to protect healthy garden soil from compaction. Through compaction, microorganisms are destroyed, their tunnels that allow air and water to infiltrate the soil, and the loose, crumbly texture of healthy soil. This is why I highly recommend building permanent garden beds and pathways, to avoid having to start from square one each year building up soil. In permanent beds, simply continue to improve their fertility year after year. This saves time, too.

Over time, the soil will become a resilient and balanced ecosystem, able to regulate and regenerate itself into the future. This doesn't mean there will be no maintenance, but anytime you can allow nature to do the work for you, the system will be stronger and less will be required of you.

Get a Soil Test

A soil test is an option that appeals to many micro-farmers because it will inform you about the nutrient content and pH level of your soil. You can also opt for a heavy metals soil analysis, which tests for contaminants. In residential areas where building practices, lead paint, and imported fill dirt may have contributed toxins to the soil, it's wise to do some reconnaissance.

THE SOIL IS THE GREAT CONNECTOR OF OUR LIVES, THE SOURCE AND DESTINATION OF ALL.

WENDELL BERRY, THE UNSETTLING OF AMERICA

While the results of a soil test can be informative, I'm a little wary of the *recommendations* that come with the test results. Often, the testing entity will recommend the application of synthetic fertilizers as the solution to manage nutrient deficiencies or pH imbalances. However, I believe the best way to improve soil, reduce soil erosion, and protect our local waterways is by adding organic material, not more chemicals.

TIP: ORDERING SOIL TESTS

Your local cooperative extension office can assist you in ordering soil tests. For a link to find the nearest office, go to www.TenthAcreFarm.com/tsmf-companion.

I prefer to amend soil the way nature does—by using biodegradable organic materials that break down quickly and are immediately bioavailable to plants. They are safe; it would be hard to over-apply them.

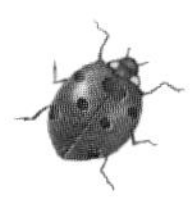

Guide to Organic Soil Amendments

Soil amendments are used in gardens for many different reasons. Some amendments improve soil health and tilth, which increases beneficial soil organisms, organic matter, and improves moisture retention. Other soil amendments add essential nutrients that help plants to grow healthy. Still other soil amendments help to balance soil pH and neutralize either acidic or alkaline soil so that soil organisms can thrive. Many soil amendments are free or fairly inexpensive, biodegradable, and easy to find locally or make yourself, whereas others are products that will need to be purchased.

In general, soil amendments are best added in the fall or spring before the garden is planted. In some regions where rainfall is more abundant and nutrients are more easily washed away, a second application of fertilizer can be helpful halfway through the season.

Reminder: Before choosing soil amendments—especially purchased ones—it is important to get a soil test to know exactly what your soil needs. The following guide will help you choose the amendments that are most appropriate for your needs.

The following symbols will help you quickly find the amendments you need.

indicates an amendment that adds or improves soil nutrients.

indicates an amendment that conditions and improves soil texture, improves drainage/moisture retention, and increases beneficial soil organisms.

indicates an amendment that raises soil pH and reduces acidity.

indicates an amendment that lowers soil pH and reduces alkalinity.

Aged leaf mold

Animal-Based Soil Amendments

Animal-based soil amendments can be a potent source of nutrients and can increase beneficial soil organisms. Some amendments can also improve soil structure and affect pH levels. Untreated animal products—such as fresh manures, bone and blood meal, and fish emulsion—are most safely applied nine months before harvest, or at a minimum of two weeks before planting.

Bat Guano: Bat guano—or bat poop—is a fast-acting, organic fertilizer high in nitrogen and phosphorus, which promotes strong and healthy plant growth. It can also improve the texture of soil, improve drainage in heavy soils, and help to neutralize soil contaminants. By increasing beneficial bacteria in the soil, it helps to protect plants against disease. Bat guano is highly concentrated, so although it can be expensive to purchase, a little will go a long way.

As with other animal manures, it is best mixed into the soil in the fall, or at least two weeks before planting, to allow time for the nutrients to break down into a form that plants can absorb. Bat guano is considered by many to be unsustainable since the harvesting of it can destroy cave habitat and may negatively affect the health of bat populations.

Blood Meal: Blood meal is a by-product of animal processing. It is one of the highest non-synthetic sources of nitrogen, and is quick-acting. Nitrogen is essential for healthy, green vegetative plant growth. Blood meal can lower soil pH making soil more acidic, and therefore shouldn't be used on soil that is already acidic.

One application per year is usually all that is needed, but be sure to follow product instructions. Because this amendment is fast-acting and can burn plants or reduce germination rates if too much is used, err on the side of adding less than product instructions. It can also be added to compost piles to balance a high carbon content. Although blood meal can deter herbaceous animals such as rabbits and deer, it may attract omnivorous animals such as raccoons and dogs.

Bone Meal: Bone meal is also a by-product of animal processing. It is a slow-release fertilizer. Be sure to know your current soil pH before using it, because it can lower soil pH. Bone meal is

not bioavailable to plants growing in soils with a pH over 7.0. It is used mainly for its high phosphorus content—important for strong root development—but it also contains some calcium and other trace minerals.

It is usually mixed into the top few inches of garden soil in the spring before planting. Mycorrhizal fungi in soil will help break down the phosphorus for the plant roots to absorb, so consider adding mycorrhizae inoculant along with the bone meal. Be aware that it can cause harm to dogs if consumed.

Eggshells: Eggshells are especially high in calcium and can be beneficial to certain crops. I prefer to compost eggshells before using them in the garden, because once broken down, the calcium will be present in the compost in a form that is highly absorbable for plants. In a study done by the Department of Agronomy & Soils at Auburn University, crushed eggshells that were added directly to the soil had no effect on soil structure or calcium levels. However, when dried and powdered first, eggshells were as effective as any store-bought calcium amendment.

For a quick soil fix, add a teaspoon of dried, powdered eggshells to planting holes or broadly mix into the soil when planting. The following crops will favor powdered eggshells in planting holes: beans, broccoli, cauliflower, cherry trees, cucumbers, lettuce, peppers, potatoes, squash, tomatoes. Because calcium can make the soil more alkaline, add eggshells only if a soil test recommends a change in pH level. Gypsum, another soil amendment, can add calcium without raising soil pH.

Fish Fertilizer: Fish fertilizer is made from the byproducts of the fish oil industry and can be a good source of nitrogen and other trace minerals. Legend has it that Native Americans and early American settlers buried fish parts in their cornfields before planting. Fish fertilizers can come in the form of fish meal (dried and powdered fish byproducts), hydrolyzed fish (fish parts broken down using enzymes into a compost tea-like product), or fish emulsion (the liquid leftover from the other processes).

Hydrolyzed fish is the most potent and most readily available form of nutrients, while fish emulsion is the least potent. They are both often sold as a concentrate that needs to be diluted as directed before using. Fish meal is used as a slow-release fertilizer that is mixed into the top few inches of soil before planting.

Fish emulsion is the product that I use the most often because it is the most affordable. I often dilute it one-half times more than the recommendation, and water the garden with it once a month. In general, all forms of fish fertilizer have slower release rates than other types of fertilizer, but over time it will improve the nutrient content of the soil. Some users report that the fish hydrolysates attract the most beneficial microbes.

Manure: Livestock manure is used mainly as a slow-release fertilizer, because it contains most of the elements required for plant growth, including nitrogen and many other nutrients. It can also condition the soil, increasing beneficial soil organisms and moisture retention. The manure can come from nearly any livestock animal. Do not use manure from dogs or cats.

I often find horse manure at local farms where it is given away for free if I'm willing to pick it up and haul it away myself. When finding livestock manure locally, look for farms that pasture-raise their animals and feed them organic feed, since manure from other types of farms can include herbicide residues that can stunt plant growth.

It is okay to spread fresh manure on the garden if you are doing so in the fall when the garden season has ended, because it will have composted by the time the spring planting season rolls around. Fresh manure should be spread at least three to four months before a crop will be harvested in order to avoid potential pathogens.

However, because fresh manure can burn plants as well, it is usually aged—or composted—with livestock bedding for six months before spreading in the spring. Aged manure will contain less nitrogen than fresh manure, but it will make an exceptional soil conditioner. It should be spread at least one month before planting crops. Aged manure will have fewer weed seeds than fresh manure, which may reduce your workload during the garden season.

Manure—whether fresh spread in the fall or aged spread in the spring—should be turned into the soil within 12 hours of the time of spreading, and certainly before spring planting. Turning manure into the soil ensures that more nitrogen is captured in the soil rather than leaching away.

When spreading fresh manure and most other soil amendments, be sure to do so on ground that isn't frozen or oversaturated by a recent rain. If a heavy rain is in the near forecast, wait it out. These are common agricultural prohibitions included in many state laws that are helping to reduce runoff from farm fields to keep our waterways clean. Plus, you don't want all of your hard work and valuable nutrients to wash away!

Oyster Shells: Crushed oyster shells are an impressive slow-release fertilizer and soil conditioner for the garden, composed of 95% calcium carbonate. They will raise pH, helping to neutralize acidic soil, loosen clay soil, and improve drainage. Ground oyster shells can be used in place of lime to improve soil pH, and they can be used in place of dried and crushed eggshells as a calcium supplement, though the calcium will be slower to release.

Powdered oyster shells can be added to the planting hole of calcium-loving crops such as tomatoes and peppers. Underground stores of crushed oyster shells will repel digging animals such as moles and voles. They can also be used on top of the soil around pest-prone plants to deter slugs. If a more immediate calcium supplement is needed, soak one part crushed oyster shells with three parts apple cider vinegar for two or three weeks. Mix ¼ cup of the vinegar solution with one gallon of water and spray the foliage of calcium-loving crops.

Worm Compost: Worm castings—or worm poop—are the richest fertilizer known to humans, made up of 50% humus. They are high in minerals as well as nitrate, a more bioavailable source of nitrogen than that found in other fertilizer sources. They help neutralize soil pH. As a bioactivator, worm castings add humic acid to the soil, which stimulates plant growth and increases microbial activity. Worm castings help plants regulate water usage, improve soil structure, and increase plant vigor. They can even be used in place of potting soil and to filter out contaminants.

Worm castings are so safe that there is no upper limit to how much you can apply at one time. I can't say enough good things about this soil amendment. See "Construct a Worm Bin" for instructions on how to start your own worm compost bin.

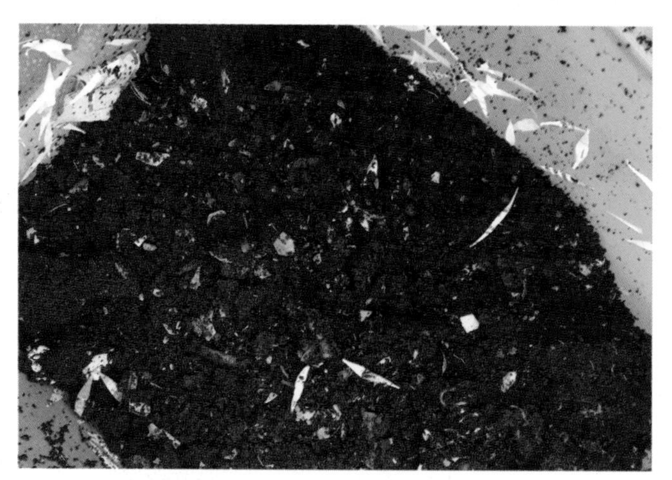

Fertilizer-rich worm castings ready for the garden

Mineral-Based Soil Amendments

Mineral-based soil amendments are most often used to correct mineral deficiencies. Some of them can also affect soil pH and structure. Because mineral-based amendments do not break down easily in the soil, it is essential to get a soil test before applying them so as not to add too much.

Epsom Salt: Epsom salt is also known as magnesium sulfate, two elements that are important for a garden in which a soil test has determined a need for them. Leaf curling and yellow leaves could be an indication of magnesium deficiency. Fans of this mineral compound swear that it improves seed germination, plant vitality and growth, and the absorption of nutrients that already exist in the soil. It is usually worked into the soil before planting. It can also be used as a foliar spray throughout the garden season, which is a more effective way to apply it in alkaline soil.

Tomatoes and peppers will reportedly benefit from Epsom salt treatments. Perennials may also benefit, showing greener foliage and sweeter fruit. Although official studies can't corroborate Epsom salt's effectiveness in the garden, its fans certainly swear by it.

Greensand: Greensand is a slow-release soil conditioner. It is largely composed of glauconite, a mineral harvested from ancient forest floors. Greensand is considered high in potassium and trace minerals such as iron and magnesium, but its main benefit is that it loosens clay soil and improves moisture retention. Greensand should not be confused with regular sand, which when mixed with clay soil, can produce a cement-like mixture. Apply in early spring before planting.

Lime: Agricultural lime is ground limestone, a naturally occurring rock that is high in calcium. Lime is used in gardens as a slow-release calcium supplement and to raise pH to make soil more alkaline. Before using lime, a soil test should indicate that you have acidic soil. Lime should be mixed into the soil in the fall on dry ground. Careful; a little bit goes a long way.

Don't add other fertilizers in the same year as lime. Test the soil and apply lime, if needed, every three years.

Rock Dust: Rock dust is a slow-release fertilizer made from mined rocks that are high in particular minerals. They can benefit soils that have been cultivated for some time and where a soil test indicates a need for a nutrient boost. For example, rock phosphate is high in phosphorus much like bone meal. Granite dust is high in potassium. Basalt dust is high in phosphorus, potassium, calcium, magnesium, and iron.

The dust should be mixed into the soil in the spring before planting according to product instructions, and should not be used more often than every three to five years.

Keep in mind that minerals are difficult for plants to access in the soil, and it takes a lot of beneficial soil organisms to break them down into a bioavailable form. For this reason, compost and other amendments that boost soil organisms should be spread at the same time as the rock dust. Not only will adding compost help rock minerals to be available to your crops, but also you'll minimize the amount of rock dust you need to apply, saving you money in the long run.

Sulfur: Sulfur is an essential nutrient for healthy plant production. Though it is difficult to test for a deficiency of this element, yellowing leaves and general lack of vigor can be indicators of the need for this slow-release fertilizer. Sulfur is needed most often in high alkaline soils, and will help to lower the pH to more moderate levels. It is easiest to improve the sulfur content of soil by adding manure regularly, because the effectiveness of the powder form of sulfur can be tricky, and too much may be detrimental. If using the sulfur dust, follow soil test results, and turn it into the soil in the fall. Other materials that can reduce soil alkalinity are pine needles, shredded oak leaf mold, and peat moss.

Plant-Based Soil Amendments

Plant-based soil amendments can be used to improve soil structure, balance soil pH, and improve nutrient content. It is important to source herbicide-free plant-based amendments in order to avoid contaminating the soil. The result of herbicide contamination is low germination rates and curled/yellowing leaves.

Alfalfa Meal: Alfalfa meal is made from fermented alfalfa plants, which have deep roots that accumulate deeply embedded minerals. It contains triacontanol, a naturally occurring growth hormone, which is a bioactivator, meaning that it triggers biological processes, helping plants to absorb the nutrients more effectively. Alfalfa meal is most often used as a source of nitrogen, calcium, smaller amounts of phosphorus, potassium, and other micronutrients that get depleted over time with garden cultivation. Alfalfa will increase the pH of acidic soil, and shouldn't be used on alkaline soil.

Mix the powder into the top few inches of garden soil in the spring just as perennials are showing new growth. It can also be added to compost piles to stimulate decomposition. If organic gardening is important to you, be sure to look for organic alfalfa, as most alfalfa is genetically modified and sprayed with herbicide.

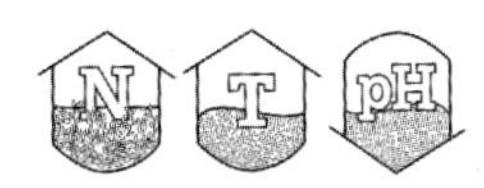

Coffee Grounds: Coffee grounds will quickly improve soil health and soil tilth. While they are a decent slow-release fertilizer of nitrogen, phosphorus, potassium, and other trace minerals, they are primarily a soil conditioner. Worms and other soil microbes will till them into the soil for you. Coffee grounds are also very acidic, which helps to balance naturally alkaline soil. The acidity and scent will repel slugs, snails, and even cats.

I pick up coffee grounds from my local Whole Foods grocery store for free. You can also check with coffee shops. A friend of mine gets her coffee grounds from a White Castle store, and in the past I have received them from a Starbucks.

The cultivation of conventionally grown coffee is one of the most chemically laden crops, applying as much as 250 pounds of fertilizer per acre, and is one of the most intensively sprayed crops with pesticides. It is unclear whether pesticide residues remain in a cup of joe or in the remaining grounds after the brewing process. However, if conscientious consumerism and a chemical-free micro-farm are important to you, you'll want to think carefully about your source.

Although coffee grounds have been traditionally recommended as a top dressing for acid-loving crops like blueberries, carrots, lettuce, and sweet potatoes, recent lab results reveal that they may be even too acidic for these plants.

Apply coffee grounds to the garden late in the fall or early in the spring when the beds are resting—at least two weeks before planting. Not much is needed; just ½ inch is plenty. Mix them in with a digging fork. Or add them to the compost bin. Be sure to know your soil's pH before adding coffee grounds directly to it. If your soil leans toward naturally acidic, composting them first will be your best bet.

Because coffee grounds are beloved by worms, I add them to my worm bin, too. See "Construct a Worm Bin" in this chapter for instructions on building your own worm composting bin!

Comfrey: Comfrey is a perennial herb with large green leaves and purple, pink, or white flowers that grows in hardiness zones 3-9. Comfrey's deep roots condition and mine the subsoil for nutrients and accumulate those nutrients in its leaves. Its nutrient levels rival those of animal-based amendments. Comfrey can be used in many ways to fertilize soil. It can activate a compost pile due to its high nitrogen content. The chopped leaves can be used as mulch around fruit trees and mature fruiting vegetable plants. Comfrey leaves can also be used as a green manure—spread on garden beds in the fall and turned under in the spring before planting.

Comfrey powder can be used to fertilize a garden bed in the spring before planting, To make a comfrey powder, simply hang the large leaves until dry, then grind them into a powder that can be sprinkled over garden soil. Be aware that comfrey leaves contain small hairs that can hurt if they pierce the skin, so wear gloves when handling. Sprinkle the powder along each row and mix into the soil with a digging fork a couple of weeks before planting.

Comfrey is cut to make mulch and to dry for making powder.

KNOW WHAT'S IN YOUR COMPOST

Purchasing commercially produced compost (sometimes called compost soil) has been touted as an ecologically friendly way to keep waste out of the waste stream while improving your garden soil. Unfortunately, it isn't quite that simple. When buying bagged compost, search the ingredients for something called "milorganite", which is a product made from the municipal sewage waste of the city of Milwaukee, Wisconsin. Now, I am all for recycling human waste in small, controlled environments such as residential composting toilets. But municipal sewage sludge—which is mixed with industrial wastewater—often contains heavy metals as well as traces of hormones and prescription drugs.

Bulk-made commercial compost can unfortunately yield similar results. Trucks deliver organic matter that is sometimes mixed with industrial waste, which could be tainted with heavy metals. Some composting operations also mix in biosolids (sewage sludge), yielding similar risks as bagged compost. When purchasing compost in any form, be sure to ask what it was made of, and whether there has been any testing to determine the presence of heavy metals or other toxins. Most of us grow some of our own food because we want to know what it is made of, and these industry solutions have only served to muddle that process.

Compost (Homemade): Homemade compost made from food scraps and yard waste is an inexpensive, slow-release fertilizer and soil conditioner for the garden. It's also a great way to keep household waste out of the waste stream. Be sure to use only compost that is completely decomposed in garden soil. Partially decomposed compost contains bacteria that will compete with vegetable plants for nutrients and substantially reduce germination rates.

As a soil conditioner, homemade compost will improve the structure of soil by aerating existing soil and improving drainage and moisture retention. While compost includes only low levels of nitrogen, phosphorus, and potassium—the three essential nutrients for plant growth—it does include a variety of micronutrients. Beneficial soil organisms and worms will assist in the breakdown and absorption of nutrients. The increase in microbial activity helps plants fight off diseases and pests.

Add three to four inches of compost to garden soil each spring before planting and work it in with a digging fork. For perennials, spread compost annually around trees and shrubs without working it into the soil. See "Build Your Own Compost Bin" for instructions on how to build your own compost bin and what to compost.

Cover Crops: Cover crops are plants that are seeded in empty garden beds in the fall, about four weeks before the frost date. The cover crops will overwinter, and by springtime they will have grown full and lush, outcompeting early spring weeds. Just when they are flowering or setting seedheads, they are cut back just above the soil line. After a couple of days, the "green manure" is incorporated into the soil with a digging fork, breaking up roots. Many micro-farmers use livestock, such as chickens, to help turn it in. Wait three weeks before planting in the bed.

Alfalfa and legumes such as field peas and clover are cover crops that will fix nitrogen in the soil, while cover crops like buckwheat and rye will add biomass. Many micro-farmers use a mixture of different species. There are many kinds of cover crops, and which mixture you use will be dependent on your local climate and your goals. If your garden is a no-till garden, avoid grass-type cover crops since they will be a challenge to hand-turn into the soil. Your local extension office can help you choose appropriately.

When cover cropping, I alternate keeping a few garden beds for overwinter vegetable production, while planting the rest in cover crops.

The main reasons for using cover crops are increasing soil fertility through nitrogen fertilizer, adding organic matter to build humus, improving soil texture, and increasing beneficial soil organisms and fungi, all which help to reduce pests and disease. Another benefit is reducing soil erosion during a time of year when soil is most often exposed to the elements. You can even consider taking a garden bed or two out of production for an entire season and allowing the flowering plants to attract beneficial insects and pollinators.

If you've purchased bulk soil for your new garden beds and have seen lackluster growing success, adding organic matter and nitrogen—which the industrial soil is typically lacking—through cover crops can help.

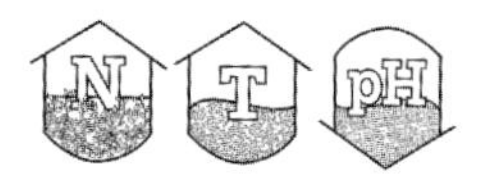

Cottonseed Meal: Cottonseed meal is a byproduct of the cotton industry and is used as a slow-release nitrogen fertilizer. It is high in other nutrients as well. It can improve soil texture, aerate heavy soil, and improve moisture retention in light soil. It will help to balance alkaline soil or it can boost acid-loving plants. It is often mixed into soil along with compost or leaf mold.

Cotton is a crop that is often genetically modified and heavily laden with pesticides, and at the time of publication, I couldn't locate any sources of organic cottonseed meal.

Herbs (fresh or dried): Herbs have long been used as potent sources of nutrients and medicinal qualities for human health. Using herbs as soil amendments can provide these same benefits to the garden. Dried or powdered herbs can be sprinkled and mixed into garden soil in the fall for overwintering, or in the spring before planting. Fresh herbs can energize a compost pile that is full of brown material such as leaves, or they can be steeped in water to make a liquid fertilizer to give a mid-season boost to perennials and mature fruiting vegetables. Fresh herbs can act as a green manure when chopped and incorporated into garden soil in the fall or at least two weeks before planting. Fresh herbs can even make effective chop-and-drop mulch. When laid on top of the soil, the chopped herbs will act as a slow-release fertilizer.

Leaf Mold: Leaf mulch that has aged for two to three years is called leaf mold, and it can benefit the garden in many ways. Its consistency lies somewhere between shredded leaves and leaves that have composted into humus. It is crumbly, has an earthy scent, and has been reported to hold up to five times its own weight in water, making it an effective, water-retaining mulch or soil conditioner. It has twice the mineral content as manure.

When the hot gardening season strikes, lay leaf mold over the garden as mulch, keeping it away from the stems of plants. It will have a cooling effect on the soil, and as the mulch breaks down over the course of the year, it will attract beneficial soil organisms while transforming into humus. This process will, over time, help to bring soil into pH balance, and will add micronutrients to the soil.

To make leaf mold, shred the leaves first by running over them with a lawn mower, or by using a leaf mulcher. To make "quick" leaf mold, make a rectangular pile of shredded leaves that is five feet square by five feet high. Turn the pile monthly, and you might be able to make leaf mold in as little as 12 months, though the process usually takes a couple of years. Alternatively, add shredded leaves to the compost bin.

As a soil conditioner, add finished leaf mold to garden soil in the fall, then mix it in with a digging fork in the spring before planting.

Mushroom Compost: Mushroom compost is the growing medium that is leftover from commercial mushroom growing operations. The compost will contain some nutrients, but it is used more as a soil conditioner to increase organic matter and microorganism activity, and improve water retention. Mushroom compost is alkalinizing, meaning that it will help to balance out acidic soils and isn't recommended for already alkaline soils. It shouldn't be used around acid-loving crops such as blueberries and other fruit crops.

Mushroom compost can contain pesticide residues leftover from the mushroom growing operation, so unless you can source it from a local mushroom grower who uses organic methods, it may not be the best solution for organic gardeners.

Another problem with mushroom compost is that it may be high in soluble salts depending on the source. These salts can negatively affect seed germination and the health of young plants. To minimize that risk, spread mushroom compost over the garden in the fall and allow it to settle over the winter. It is then turned into the soil before spring planting. Mushroom compost can also be spread as mulch around perennials, which can more easily accommodate the salts.

Peat Moss: Peat moss is decomposed sphagnum peat moss that comes from bogs in northern latitudes, most notably, Canada. Its fluffy, light texture improves aeration, drainage, and composition of heavy soils. Peat moss will also make soil more acidic. It is used as a soil conditioner rather than a fertilizer because it contains few nutrients. For this reason, it is usually combined with compost or manure when adding it to the garden.

Peat moss as a product is in the environmental hot seat because it takes thousands of years to form. If this is a concern for you, bone meal, coffee grounds, leaf mold, and worm compost are all good alternatives to make soil more acidic, and compost will help improve soil composition and drainage.

Seaweed Fertilizer: Seaweed fertilizer can come as a meal, powder, or liquid concentrate. It is high in potassium, a variety of trace elements, and plant growth stimulants that help to improve plant vigor. Research has documented larger yields, less susceptibility to pests and disease, better seed germination, and better frost tolerance in plants that were given seaweed fertilizer. Because seaweed has a rapid growth rate, it is a renewable resource.

Seaweed meal or powder is mixed into the soil prior to planting, while liquid concentrates are diluted as a foliar spray, are more fast-acting, and are used throughout the season. Because it is considered more of a soil conditioner than a fertilizer, it can be used in conjunction with other fertilizers that do not have a high potassium content.

Soybean Meal: Soybean meal is a plant source of nitrogen fertilizer that also lowers soil pH. If a soil test determines nitrogen is lacking, use this product sparingly, since too much can burn plants. It is mixed into the soil at least two weeks before planting. Since most soy is genetically modified, seek out organic products. It should not be used in garden beds where seeds will be directly sown, as germination rates could be reduced.

Wood Ash: Hardwood ash from wood fires and wood burning stoves is commonly used for raising soil pH in place of lime, and it also contains a variety of nutrients including potassium and phosphorus. It can even help plants regulate water.

Be sure to use actual wood ash, not charcoal briquettes or coal ash. Just a sprinkling will do—too much wood ash can negatively affect soil health. It is best applied in the fall where you intend to plant any of the following in the spring: broccoli, collards, beans and peas, root vegetables, and tomatoes. They will enjoy this treatment if the soil pH is appropriate. Apples and other soft fruit bushes will enjoy small sprinklings of wood ash in the soil underneath them in the fall.

As with all powdered soil amendments, it is a good idea to wear a mask and gloves while working with wood ash.

I prefer to add wood ash in small amounts to the compost pile (a sprinkling each time I add new material) and feed my garden through the finished compost. This way, I can take advantage of the nutrients in wood ash while preventing it from affecting my soil pH. Keep wood ash dry until you are ready to add it to the pile.

Wood Chips: With the growing popularity of the film *Back to Eden*, wood chips are being added to gardens at an accelerated rate. However, it is important to know how to use this soil amendment

correctly. Wood chips are beneficial to the garden because as they break down, they create humus, the organic component of soil that is necessary for plant life.

Adding wood chips is like mimicking the forest floor, where leaves and twigs naturally decompose on top of the soil. Wood chips improve the nutrient levels of the existing soil as they break down, and they increase the numbers of beneficial soil organisms. They hold in moisture, reducing irrigation needs. Covering the ground, they reduce weeds. Wood chips create a stable growing environment by insulating against the hot summer sun and freezing winters.

Wood chips were meant to be used as mulch, not tilled into the existing soil. Tilling them in causes them to bind to nitrogen in the soil and temporarily prevent it from being available to your plants. To use wood chips in the vegetable or perennial garden, age them for two or three years before mixing them into garden soil as an amendment, and add a teensy amount of blood meal with them to make up for lower nitrogen availability. Or lay fresh wood chips on top of the soil as mulch without mixing them in.

COULD HERBICIDES BE POISONING YOUR ORGANIC SOIL AMENDMENTS?

For as long as humans have grown food in permanent agricultural settlements, organic amendments such as livestock manure and plant residue have been used to boost the health and productivity of garden beds and farm fields. Unfortunately, an herbicide called aminopyralid was released to the public in 2005, which now persists in the very organic amendments that we think of as pure and safe. The general term for amendments that have been poisoned by herbicides is "killer compost".

The herbicide, when sprayed in farm fields to reduce weeds, tends to be persistent, which means that it persists in the soil and plant residue. For example, a horse may eat hay that was sprayed with herbicide in the field. The herbicide will persist through the digestive system of the horse and continue to be present in the manure. When the manure is spread on the garden, it will pass along the herbicide to the soil, resulting in stunted growth and lack of vigor in garden plants. Straw is another organic amendment that is commonly laced with herbicides in the same way.

According to The Ohio State University, these poisons are so powerful that they can damage crops at levels as low as 10 parts per billion. The label of the product specifically says to never use the treated materials or manure as a soil amendment. The trouble is that the product is widely used, and farmers haven't been educated properly about its potency.

The moral of this sad story is that it is essential to know the source of your amendments. If you can't make it yourself, try to find it from a local farmer or producer who can answer your questions about how it was made and the chemical controls that may have been used in the process.

Wood chips are a free or low-cost renewable resource for the garden.

I like to use wood chips in the pathways of my vegetable garden, where I can reap some of their benefits in the beds themselves. (Beneficial soil organisms and fungi will enjoy plowing through your beds as they create connections between the pathways.) Tree trimmers will often deliver wood chips for free if they are working in your area. I have access to an arborist who delivers a trailer-load for a $20 fee.

Choosing a Soil Amendment

Most soils—whether due to ancient land formations or the over-cultivation of a growing area—will need more of certain nutrients in order to be ideal growing spaces. Soil testing will help identify exactly what your soil needs.

The primary nutrients of nitrogen (N), phosphorus (P), and potassium (K) are what plants use large amounts of for growth. Therefore, amending the soil with these nutrients is essential. Secondary nutrients such as calcium (Ca), magnesium (Mg), and Sulfur (S) are also important, and many soils will have sufficient amounts of these. A group of micronutrients, including copper (Cu) and iron (Fe), are also essential for plant growth but only in very small quantities.

Conventional modern farming practices tend to be hyper-focused on synthetic versions of the primary nutrients, N-P-K. The trouble is, even with sophisticated measuring equations, the absorption rates of these synthetic nutrients are astoundingly low, and rain events send massive waves of unused nutrient and polluted topsoil into our waterways.

Case in point is the Gulf of Mexico "Dead Zone," which measured 5,052 square miles in 2014. According to the National Oceanic and Atmospheric Administration (NOAA), a dead zone is caused by "excessive nutrient pollution from human activities coupled with other factors that deplete the oxygen required to support most marine life in bottom and near-bottom water." An increase in chemical nutrients is one culprit, as is polluted runoff from cities.

One of the missing links here is the connection between soil ecology and absorption.

You can maximize nutrient absorption rates—and reduce pollution—by improving the microbial activity of the soil. Many nutrients added as soil amendment products aren't in a form that is usable by plant roots. Rather, beneficial soil organisms get to work transforming the amendments into usable resources. Where there is more soil life, there is more bioavailability of the administered nutrients. In the end, focusing more on increasing healthy soil life will help maximize the small amounts of fertilizers added.

So how do you create a soil environment that is teeming with microbes? The best way to do it is by adding organic matter, and the best way to do that is by using the organic matter that you have available to you for free—homemade compost, worm castings, manure, leaf mold, and coffee grounds, to name a few. Others, such as bat guano, fish fertilizer, and cover crops, will also help to benefit soil microbes, and are reasonably inexpensive to purchase.

This means that when using other purchased amendments, you can start below the recommended application while continually adding natural and homemade soil conditioners to make them more bioavailable to plants.

In the end, if you focus at least as much—and I would encourage more—on adding organic matter and increasing beneficial soil organisms, as you do on the purchase of fancy soil amendments, you'll maximize the efficiency of your efforts and reduce your micro-farming costs, too. You'll also reduce the chance of sending excess nutrients into your local waterways.

Try it yourself and see what works for you.

 No-Till Micro-Farming

Industrial farming on large acreage requires tilling to loosen the soil. Unfortunately, this loosening destroys beneficial soil organisms while allowing the soil to blow away in the wind and wash away in the rain. According to the USDA Natural Resources Conservation Service, America loses almost *three tons of topsoil per acre per year*, making it the country's largest export. Since topsoil is required for growing crops and can take up to 500 years to form naturally, this is alarming news.

Fortunately, growing crops at the micro-farm scale doesn't usually require tilling to produce a loose soil for planting. A digging fork is a useful tool for the no-till garden because it will easily loosen the soil without destroying the soil microorganisms. Did you know that earthworms, fungi, and other beneficial soil organisms exude sticky substances as they build tunnels and networks throughout the soil? These sticky substances are nature's way of holding soil together so it doesn't wash or blow away even when it is loose and crumbly.

This is the reason why aerating and loosening the soil with a rototiller is different than using a digging fork: A rototiller chops up worms and other soil organisms, destroys the fungal networks and tunnels made by soil organisms, and disrupts the connections of sticky exudates. This activity kills soil fertility, requiring fertility to be added in the form of imported soil amendments. What is ignored in this scenario is that fertility isn't solely about nutrients; it's also about beneficial organisms that make the nutrients bioavailable to plant roots.

The digging fork loosens and aerates without destroying soil life.

Rather, a digging fork gently enhances the top few inches of soil disturbed by annual gardening activities such as planting, weeding, and harvesting.

Transitioning to a No-Till Garden

For seasoned micro-farmers who might wish to abandon the tiller for a no-till garden, there are a few habits that can make the transition a successful one. Alas, it isn't as easy as just putting away the tiller. Leaving the tiller behind yields one glaring problem in the short-term: Weeds.

Although weeds are common in the first year post-tilling, it is no less stressful. When weeds exist in heavy numbers, they are trying to do the job that nature has set out for them: fertilize the soil. Now, to be sure, there will always be weeds to some extent. But an overabundance of weeds is an indication that the soil lacks organic matter and needs nutrition. If you can stomach it, let the garden go for a year. Check on the garden once a week, chopping and dropping the weeds. Chopping and dropping regularly is very important! Otherwise, the weeds will set seeds everywhere. The chopped-and-dropped weeds will fertilize the soil as they decompose, and their decaying roots will feed beneficial soil organisms.

> WE KNOW MORE ABOUT THE MOVEMENT OF CELESTIAL BODIES THAN ABOUT THE SOIL UNDERFOOT.
>
> *LEONARDO DA VINCI*

After weeding a no-till garden, mulch heavily with chopped herbs of all kinds, as they will also fertilize the soil. Comfrey is a good example, but other culinary and medicinal herbs work well, too.

If letting the garden go like this makes you uneasy, consider growing a cover crop for the season instead. Cover crops will not only protect the soil as mulch, they will also fertilize the soil with nitrogen and other micronutrients, as well as attract soil organisms. Choose non-grass cover crops for the no-till garden, since grasses need to be tilled under and are not well-suited to the hand-tended garden. For more ideas, find "Cover Crops" in the "Guide to Organic Soil Amendments" section earlier in this chapter. This is not essential every year, but it can help improve the health of your soil after years of tilling.

After a year of allowing weeds, cover crops, and herbs to fertilize the soil, a soil test can determine which nutrients your soil is still lacking. Store-bought soil amendments may be the final step before planting. It's important to note that soil amendments are digested by soil organisms, which then make the nutrients available to plants. The more soil organisms you attract, the fewer amendments you need to purchase because you'll have a higher rate of absorption.

All of these activities will help attract soil organisms and help to balance things out so that weeds are fewer in years to come.

Although you want to leave the soil intact as much as possible, it is common for the top few inches of soil to be gently disturbed for weeding, planting, and harvesting. This activity allows a bit of aeration without destroying soil organism habitat or beneficial fungal networks. When you're

ready to plant in the bed, use a digging fork to poke holes throughout the garden bed to loosen the soil, improve drainage, and turn over the weeds or cover crops gently.

Keep in mind that the texture of no-till garden soil will be different than that of a tilled bed. It will be denser, like chocolate cake, which everybody likes! In time, your no-till-grown plants—with the help of fungal networks—will better regulate water usage, making them more resilient through super wet or dry periods.

In the long term, there are a few habits that can help maintain healthy soil in the no-till garden. The first is by crop rotation. Through tilling, soil diseases and pests are exposed to the air and elements, thereby reducing their viability. In no-till gardens, however, where the soil is not disturbed regularly, diseases and pests can persist. To combat this, keep good garden notes and rotate your crops each year, especially after a pest or disease outbreak. Your soil will already be healthier with increased numbers of beneficial soil organisms to promote overall health and the digestion of nutrients. Healthier plants are less susceptible to diseases and pests.

Winter cover crops, as mentioned above, will protect the soil, fertilize, and attract beneficial soil organisms.

To protect the soil in a no-till garden, don't walk in the beds. While in the tilled garden the soil is loosened every year mechanically, the only way to keep soil loose for planting in the no-till garden is by avoiding compaction. Compaction destroys beneficial soil organisms, their tunnels that allow air and water to infiltrate the soil, and the naturally loose, crumbly texture of healthy soil.

Another way to protect garden soil is by adding organic matter—lots of it, as often as you can. To read more about what kinds of organic matter to add, see the previous section, Guide to Organic Soil Amendments.

TIP: STAY OFF THE BED!

To keep from walking in the beds, build permanent garden beds and pathways to avoid having to start from square one each year building up soil. In permanent beds, simply continue to improve their fertility year after year. This saves time, too.

Mulching can also help to protect soil. Mulching shades the soil (habitat for the beneficial soil organisms) and reduces the germination of weed seeds. I mulch between the rows of veggies, taking care to avoid touching the stems of plants. I use grass clippings and leaf mulch because that's what I have the most of. Wood chips are an excellent topping. Just be sure to keep them on

top of the soil as mulch and away from plant stems. Do not mix them into the soil as an amendment, since this can reduce your garden's vigor. In the winter, try mulching with shredded leaves. Lighter mulches are beneficial in wet areas, while heavier mulches are more beneficial in dry areas. See chapter 9 for more tips on mulching.

Although the no-till garden is simply trading one activity (tilling) for another (supporting healthy soil), I believe you will get more joy and satisfaction out of learning how to fertilize, support, and protect your soil. You will also get healthier produce, which is what most of us are aiming for.

Starting a New No-Till Garden in the Fall: Sheet Mulching

Sheet mulching is a simple but effective method for creating new garden beds without using a tiller. Rather than tilling down into the existing soil, sheet mulching builds on top of it by layering new "sheets" of organic matter. To use this method, start early in the year by collecting large sheets of cardboard in addition to chemical-free organic materials like compost, worm castings, grass clippings, shredded leaves, shredded newspaper, aged manure, or wood chips that have aged for at least one year.

Sheet mulching includes laying down cardboard and topping it with organic matter. Here, I am spreading a layer of manure.

When fall rolls around, identify the area that will be sheet mulched, such as over a grassy area, inside a new raised bed, or over a garden area that has been overtaken with weeds. Use a digging fork to poke holes and aerate the hardpan soil.

Lay cardboard over the area, overlapping the ends of the cardboard sheets by six inches so the weeds and grass can't grow through. The cardboard will not only kill the grass and weeds, but it will also encourage the worms and other soil microbes to infiltrate the new garden bed. Worm activity at the root level of your plants is crucial for their health in the no-till garden.

Next, add six to 24 inches of compost soil. Layers of compost soil can be alternated with layers of free resources such as grass clippings, fallen leaves, shredded newspaper, aged manure, or wood chips that have aged for at least one year. This layering is why sheet mulching is sometimes referred to as lasagna gardening. The final top layer should be at least six inches of compost soil.

Watering it well will notify the soil microorganisms to get to work mixing it all together for you. Chlorinated water from municipal water sources can damage soil microorganisms and take them longer to build up healthy populations. For this reason, watering with rainwater can be beneficial to your new garden bed. For ideas on catching rainwater in the landscape, see chapter 11.

Let your newly sheet-mulched garden bed sit for at least two weeks before planting, ideally for three months. In time, the soil organisms will help break down the "sheets", churn the new soil amendments into the existing soil, and begin to turn the hardpan soil soft and chocolate cake-like. Add more organic matter as often as you can, especially during the first few years.

Starting a New No-Till Garden in the Spring/Summer

Ironically, tilling can help get a new garden bed started quickly when you don't have a storehouse of organic materials for sheet mulching. While I try to avoid tilling at all costs because of its damage to soil organisms and their habitat, if you really want to, a first pass with the tiller will be fine in the long run. Follow tillage immediately with all of the practices mentioned above, especially adding as much of your homemade compost, worm castings, and other organic matter as you possibly can.

For both methods, it is important to remember that the main focus is on attracting, supporting, and protecting beneficial soil organisms. This process won't happen overnight. The first year of a new garden bed rarely produces abundantly or is pest or disease free because the soil organisms aren't settled yet in the new home you've created for them. Continue to support the soil organisms, and you will be on your way to having a successful garden in the long run.

Sheet Mulching

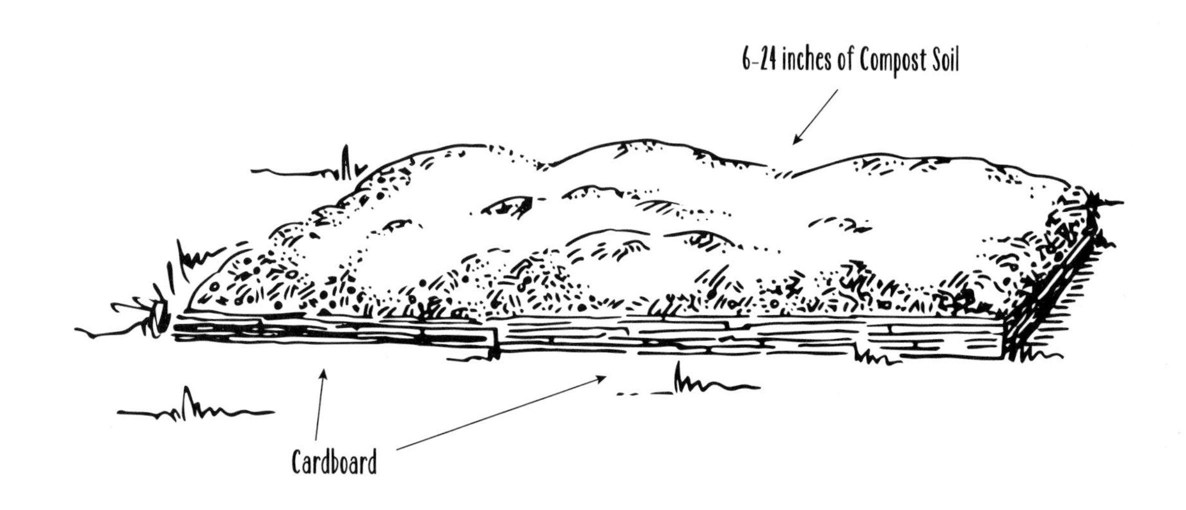

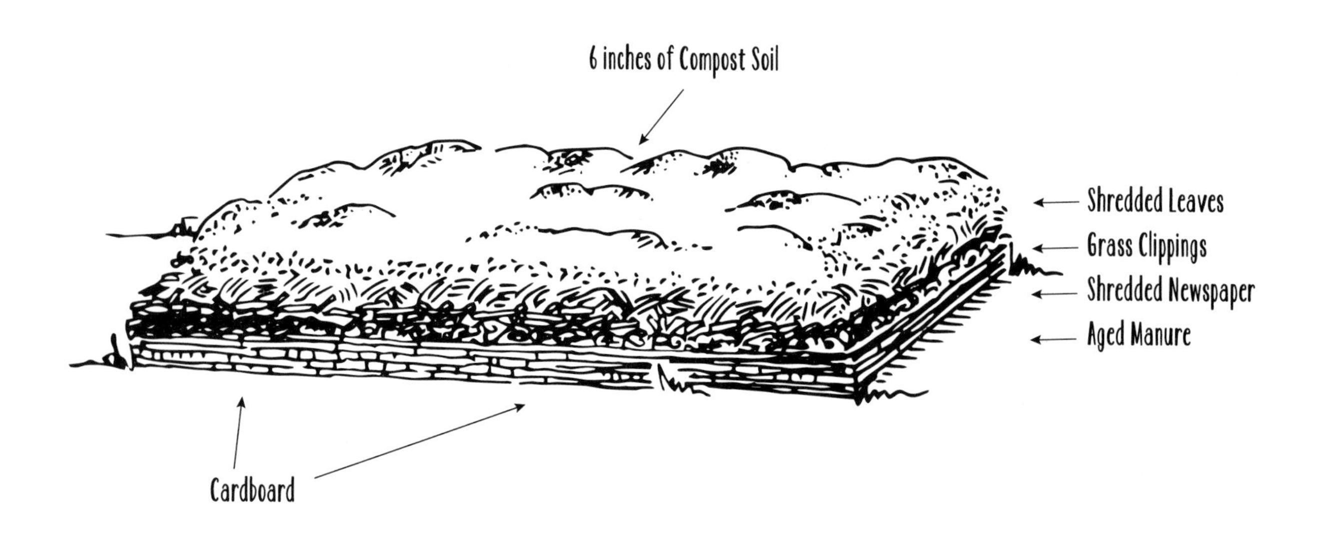

Preventing Soil Erosion

The epicenter of soil erosion damage is the midwestern Corn Belt, where corn and soy crops are grown on hundreds of acres of mono-cultured industrial farms. Topsoil blows away in the hot, dry summer, and flushes down the Mississippi River to the Gulf of Mexico where it smothers aquatic life at an alarming rate. Lost forever. Experts estimate that with current farming practices, we only have 60 years of topsoil left.

A NATION THAT DESTROYS ITS SOILS, DESTROYS ITSELF.

FRANKLIN D. ROOSEVELT, 1937

Topsoil is where soil microbes develop mutually beneficial relationships with plant roots. It's where bacteria and fungi help to hold the soil together. It's where the essential nutrients are that feed plants. Simply put, topsoil is essential to food security. Caring for the soil is crucial to the long-term viability of our micro-farms and an important consideration for developing diverse, ecological landscapes. Here are some tips for preventing erosion.

Grow Your Own Produce

Yes, having a garden is the best way to prevent soil erosion! When we become more efficient at utilizing small spaces, the less we need gigantic, clear-cut areas for industrial farm production. Returning depleted farmland back to prairie, forest, and sustainably managed perennial systems means saving what's left of the topsoil while at the same time regenerating more.

Avoid Tilling in Small Spaces

Tilling destroys and interrupts the soil life that would have helped to hold soil together and feed plant roots. In small spaces, simply loosen the garden soil each spring with a digging fork before planting.

Manage Water

In bare, loose soil, gullies form during heavy rains and wash away topsoil. In our backyard gardens and micro-farms, we can minimize water erosion by designing our gardens properly.

Build garden beds on elevation contours to catch the water and spread it across the landscape rather than sending it away. This allows the water to slowly sink in over time, irrigating the gardens and recharging groundwater. For more details on how to use the contour of your land to capture and manage water, see chapter 11.

Mulch

Mulch holds topsoil in place during rain events and keeps it moist and protected from the sun. More mulch is needed in dry regions, while less mulch is needed in wet regions. For more details on mulching, see chapter 9.

Practice Root Partitioning

Often when planting two species next to one another, whether it's carrots next to lettuce or apple trees next to strawberries, we think about situating them for proper sun exposure. For example, taller apple trees would be placed on the northern side of strawberries in order to keep from shading out the smaller strawberry plants. This partitioning of species for maximum solar collection allows you to grow a more efficient and productive landscape.

This gooseberry planting is on a steep slope, helping to slow runoff from the parking lot above it. Fruit-bearing perennials can be grown in potentially contaminated soils because heavy metal toxins do not enter the reproductive parts of plants (fruits, seeds).

This practice also extends below the soil, something that is not often talked about. The roots of plants come in different shapes and sizes, and partitioning plants by their root structure can slow down water and increase soil stabilization, thereby reducing erosion. These mixed plantings will also enhance biodiversity and improve nutrient uptake for the plants themselves.

Carrots have deep taproots, while lettuce has shallower, more fibrous roots. For this reason, these two crops can be planted closer together, maximizing productivity, but also maximizing soil stability. Carrots hold the deeper layers of topsoil in place, while lettuce holds the shallower layers.

The same is true for apple trees and strawberries. Apple trees have extremely shallow roots extending five to six feet out from the trunk at maturity, holding the top few inches of topsoil in place. Meanwhile, mature strawberries have fibrous roots that extend up to three feet deep and beyond, helping to slow down water and hold deeper layers of topsoil in place.

Use Perennials

In many regions, winter and spring rains bring the bulk of precipitation for the year, which can sometimes come in the form of heavy downpours. These heavy rains can cause major erosion if left unchecked.

Fruit trees, nut trees, berry bushes, hazelnut bushes, strawberries, asparagus, and herbs are all examples of edible perennials whose roots will help to hold the soil in place during a time of year when annual crops aren't planted. Plant these perennials as high in the landscape as possible to slow down the water.

Plant Daffodils

Daffodils dazzle the early season with their colorful flowers, but you might be surprised that their benefit to the ecosystem extends beyond beauty. Daffodils are one of the few plants—called spring ephemerals—that are active during the early spring while other plants are still dormant.

Daffodils prevent soil erosion.

As spring rains wash over a field or garden, they take topsoil and organic matter with them. Daffodils—with their root systems being active during this time—can catch and hold the moisture and important nutrients and prevent them from leaching away. For more details on how to use daffodils to prevent soil erosion, see chapter 11.

Constructing a Simple Compost System

Compost is a way to transform food scraps and yard waste—items that might otherwise go to the landfill—into a useful and free soil amendment. There are many ways to compost, but here are a few composting styles to meet the needs of beginners and serious micro-farmers alike.

Build a Compost Heap in a Day

Growing up, I learned how to compost by using a heap. We collected kitchen scraps in a compost pail, and my brothers and I took turns emptying it into a compost heap in the corner of the yard. This heap is also where we put grass clippings and other yard waste. The problem with the heap is that there is no easy way to access finished compost without disturbing the rest of the pile, so—for us at least—all of that goodness never got used. The pile just got higher!

However, there is one way that the heap method can be an extremely efficient way to make compost. If you have access to composting materials in bulk, building a heap all at once can yield lots of finished compost all at once, which is essential for anyone starting new gardens on a budget.

John Jeavons in *How to Grow More Vegetables* recommends finding a spot underneath an oak tree or other deciduous tree because they provide shade throughout the summer as well as a windbreak.

To build a compost heap in a day, measure out a square that is a minimum of four feet long by four feet wide. Outline it with temporary fencing if desired. Loosen the soil using a digging fork. The pile will measure four feet tall. Create compost quickly and evenly by collecting at least three different types of materials, plus a bit of soil, which will serve to inoculate the compost with beneficial soil microbes. Each layer will be about two inches thick and alternated in the following order, with the final top layer being soil.

The first layer will include small sticks, twigs, or dried stalks. The second layer will consist of dry vegetation such as leaves, chemical-free straw, or dried grass. The third layer will consist of green vegetation such as weeds, grass clippings, food scraps, and even coffee grounds. And of course, the final layer is soil. This pile will not require turning, and should be ready in two or three months' time.

Round Wire Compost Bins

When Mr. Weekend Warrior and I began composting together as newlyweds (romantic, right?), we started with a heap for sticks and brush, and created a simple, round bin with 16-gauge galvanized

wire fencing to compost kitchen scraps, grass clippings, and leaves. The organic material was added on top and covered with leaves.

If you're new to composting, this bin is a perfect way to get started. For the best use of the wire bin technique, I recommend starting with two holding units. These units are inexpensive and easy to manage, which are important considerations while you learn how to compost. The following design was our first system, and though we've expanded our composting operation since then, we still use these wire bins.

Arrange your two bins side by side. One will be the leaf bin, and the other will be the compost bin. Fill the leaf bin with leaves in the fall. Shredding them first with a lawn mower or mulcher is even better. Straw or other dry, brown material will work if you don't have access to leaves. Try to find organic straw, since herbicide residues in conventional straw can be persistent and reduce germination rates and vigor in plants.

To start, add six inches of leaves, straw, or even sticks and twigs to the bottom of the empty compost bin. This will allow good air circulation for faster composting. Each time you add food scraps, grass clippings, or even livestock manure (not pet manure) to the compost bin, cover it completely with a layer of dry, brown material from the leaf bin. This will prevent attracting flies and other wildlife, prevent foul odors, and speed up the composting process.

When you're ready to harvest finished compost from the wire bin, simply knock it over, harvest the finished compost at the bottom (leaving the uncomposted contents), and straighten it back up again to continue composting from the top. Once you get serious about composting, the wire bin system might not be big enough for your needs, but it is an excellent place to start. It's easy to add more wire bins to the system if you have the space for them, or try the 3-bin turning unit.

The 3-Bin Turning Unit

At some point, Mr. Weekend Warrior completed a Master Composter course, in which the 3-bin compost system was esteemed as the mother of all backyard compost bins. They look neat and orderly, and the frequent turning helps the contents compost faster. This technique will work well if you have limited space for compost bins and heaps. It is also easy to harvest the finished compost from this unit.

The way the 3-bin unit works is that food scraps are added to the bin on the left. To the side of the 3-bin unit we keep a round wire bin (our old compost bin!) filled with leaves, and whenever we add food scraps, we cover them with leaves.

BUILD YOUR OWN COMPOST BIN

To build two wire bins, you'll need:

- (2) 10-foot lengths of ½-inch wide hardware cloth
- work gloves
- wire snips
- metal file
- heavy wire for ties
- pliers

1. Trim each end of the hardware cloth back to a cross wire so there aren't any sharp edges to poke or snag hands. File the cut edge so it's safe to handle when opening and closing the bin.
2. Bend the hardware cloth into a circle, and place it where your compost will be located.
3. Cut the heavy wire into lengths for ties and, using pliers, attach the two ends of the hardware cloth together with the ties to hold the hardware cloth in a round shape.

Some people use straw or other types of "brown" material when leaves aren't available. Whatever you use, it's important to cover food scraps in order to reduce pests and odors, as well as to speed the composting process by balancing the carbon to nitrogen ratios.

Once the left bin is full, we use a pitchfork to scoop it into the middle bin. This turning process aerates the pile, helping to speed up composting.

Then we start over by again adding food scraps and yard waste to the left bin. When it's full again and ready to be scooped to the middle, the existing contents of the middle bin are moved to the bin on the right. Contents will generally take about two or three months

This round wire bin is filled with fall leaves.

to compost completely, longer in cooler weather. By this time, the contents in the right bin should be finished composting and ready to be spread in the garden. If I find any sticks or uncomposted scraps, I just toss them back into the middle bin to compost some more.

TIP: WHERE TO PLACE A COMPOST BIN

Whether you're using round wire bins or a 3-bin unit, placing your compost bin in the right location will help you have a well-functioning operation. Place them on a level, well-drained spot over soil or lawn (not a paved area). A partially shady area is ideal—such as under a deciduous tree--where the bin will have protection from the blazing, midday summer sun as well as from freezing winter winds.

We built a wood-and-wire unit, plans for which can be found at most cooperative extension websites. Wooden slats on the front of each bin are removable when it's time to scoop. The wood and wire 3-bin unit is a nice setup that helps to keep a composting operation looking nice and orderly in the urban or suburban backyard. But this design is not the cheapest to build.

At our community garden, we needed a cheaper option, so we built the same 3-bin unit using free pallets and securing the corners with T-posts. It's not as pretty, but it actually gives us a little more flexibility. To make it, you'll need seven pallets, plus an optional three more to make doors, if desired. Pallets can often be found for free at independent garden nurseries or hardware stores. Look for pallets with the HT stamp on them, which means that they were heat treated, rather than the CT stamp for chemically treated. Pallets are usually 4′ x 3.4′.

Other materials:

- 6 (6-foot) steel T-posts, plus 3 for optional doors
- heavy duty twine
- sledgehammer

Stand up two pallets to create a 90-degree L-shape and mark the ground on the outside corner where the two pallets meet. Now, set the pallets back down and use the sledgehammer to drive in a T-post on the mark. Stand up both pallets one at a time and lash them to the post with twine at the top and bottom. Attach a second pallet to the other side at a 90-degree angle to create a three-sided box, using a T-post and twine.

Repeat this process until you have three, three-sided boxes. When it's time to begin adding material to the unit, having a door on the front may help you contain more material. To make a

door, simply use a T-post on one side to create a "hinged" door with an additional pallet, which can be lashed with twine to close on the other side if desired.

The benefit of this design is that it is a temporary structure that can easily be moved and modified. When you're ready to move a pile from one bin to the next, for example, the doors can be removed for easy scooping and reattached later with more twine. Some gardeners even temporarily remove the inner wall pallet when moving piles to avoid having to scoop up and over the divider wall. In fact, this design even gives you the flexibility of moving the entire unit if needed, for very little cost. For a more permanent 3-bin unit, use 16d galvanized nails to attach the pallet corners to one another at the top and bottom.

 How to Compost

Composting is a simple procedure once you get the hang of it. Simply layer nitrogen materials such as food scraps, grass, and manure with dry carbon materials such as leaves and straw to create an environment for biological decomposition.

WHAT TO COMPOST

- Coffee grounds
- Coffee filters
- Eggshells
- Fruit and vegetable scraps
- Tea leaves + tea bags
- Yard waste (grass, leaves, weeds)
- Straw
- Livestock manure

How to Use Finished Compost

The spring is a good time to harvest finished compost and spread it on garden beds and perennials before planting time. However, finished compost can be spread at any time by following these tips.

Shovel the finished compost into a wheelbarrow, returning uncomposted food scraps or yard waste back to the bin to compost further. Now, use the finished compost in your vegetable garden beds by spreading the compost one to two inches thick and mixing it into the top six inches of soil using a digging fork. If the beds are already planted and it isn't possible to mix the compost into the soil right away, then pile the compost in a shady spot with good drainage and add it to the garden at a later time. Alternatively, you can spread compost between rows of crops, being very careful to not touch the stems of plants.

Preventing Foul Odors in Compost

Occasionally, a compost bin will have a foul odor, which is an indication that the aerobic composting process has slowed. The most common source of this problem is too much nitrogen. To correct this, simply add more dry, brown material.

Foul odors could also be a sign of too much moisture. To correct this, a non-permeable or semi-permeable cover can be placed over the bin to reduce the amount of rain that infiltrates. It will be important for the pile to stay moist, however, so lift the cover at least once a week to allow some rain to percolate through. Chlorinated municipal water is not recommended, as it can disrupt the biological process.

Too much moisture could also be an indication that the bin is not located in the proper space. Perhaps the bin is in a low spot that collects too much standing water. Placing it under a deciduous tree will shield it from heavy rains while still allowing some rain to enter.

Animal Problems in the Compost

Rodents and raccoons can sometimes be a concern with backyard composting, and rats are a common complaint among urban gardeners. When I first began composting in our backyard, I wondered if the opossums and raccoons would be attracted to our bins, and if there were a way to discourage them.

I discovered that composting only those items that are approved for backyard composting is essential to deterring animals. That means animal products such as meat, dairy, or oils should never be added to the bin. Secondly, all food scraps should be buried under a brown, carbon-based material. Above, I describe how we keep a bin filled with leaves or straw next to the compost bin,

and whenever food scraps are added, we cover them completely with leaves. With these two practices alone, we haven't had any unwanted visits from the raccoons or opossums.

Other deterrents:

- Prevent burrowing and access to the bin: Attach pieces of ½-inch hardware cloth to the bottom and over the top to prevent access. This makes it more difficult for you to access, but if you're having rodent trouble, this will be an essential step. Rats can chew through plastic, so be sure to use the hardware cloth.
- Turn the compost at least once a month to discourage them from making a permanent home in it.
- Plant mint plants around the bins to repel them. I like mountain mint (*Pycnanthemum muticum*) because it doesn't spread like other mints from the *Mentha* family.
- These problem animals do not like people, so place your bin where there is a lot of human activity and where you can visit it daily.
- Dogs and cats: I grew up composting in our backyard near the woods, where you might guess would be a prime location to get unwelcome rodent and raccoon visitors. But we never had any problems, and I suspect it had something to do with regular patrols from our dog and cat.

If you do all of these things and still have a nuisance problem, then I would suggest only composting yard waste (no food waste) in your backyard bin. Start a worm bin inside to compost food scraps.

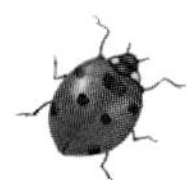

Cultivate Worm Castings with Vermicomposting

As mentioned in the above section on soil amendments, worm castings—or worm manure—are the richest known natural fertilizer. They contain an impressive list of minerals and nutrients in quantities that outperform other organic materials and manures. The phosphate, nitrogen, and potash levels in worm castings are through the roof, and all of these are essential to plant growth.

Livestock manures are important soil amendments because of the volume of manure the animals produce compared to the small bodies of worms. However, livestock manures can burn plants if added in excess, or if added before the composting process is complete. They are also not immediately bioavailable to plants.

With worm castings, however, neither of these concerns are a risk. The nutrients in worm castings are immediately available to plants, and there is no upper limit to the quantity of worm castings you can safely apply—though even a tablespoon of castings per plant would be enough to improve plant health and vigor.

That's because worm castings contain 50% more humus than regular garden soil, which is the organic component of soil necessary for plant growth and for increasing beneficial soil microorganisms. Worm castings help plants grow vigorously and help protect them from disease by forming beneficial relationships with the roots of plants. They also help retain moisture in the soil. Worm castings can be easily purchased in bags from your local garden supply store, but it's just as easy to construct your own worm bin and make your own.

If you use composted livestock manure as a soil amendment, mixing it with worm castings can skyrocket its nutrient content and its bioavailability. We don't keep livestock at Tenth Acre Farm, but we have a regular composting system that yields finished compost as a soil conditioner. We also occasionally get horse manure from a local horse stable. We mix both of these soil conditioners with our very own worm castings to make the ultimate fertilizer.

Vermicomposting is a fancy word for worm composting. It is popularly promoted as a way for space-challenged city dwellers to dispose of kitchen scraps because of the small footprint of a worm bin. However, I think worm composting is a boon for all gardeners because worm bins are so inexpensive, easy to make, and yield such an important soil amendment—for free!

TIP: WORM BIN COMPOSTING

Worm bins are an easy way to continue composting throughout the winter when the regular compost bin freezes up. I keep my worm bin in the garage, and when the compost pile outside is frozen in winter, we can continue to compost our food scraps by adding them to the worm bin.

Worm Bin Care

Feed the worms about one cup of food scraps each week for the first few months. (Remember to avoid feeding them citrus fruits, hot peppers, garlic, or onions.) As the worms look bigger and more numerous, you can increase the amount of food scraps you give them each week.

Construct a Worm Bin

Materials

- 2 (20-gallon) plastic totes + lids—dark colored, no clear plastic
- Drill with 1/4" and 1/16" drill bits
- Shredded office paper and newspaper— enough to fill half of one bin (no glossy paper)
- Spray bottle with water
- Red Wiggler worms (1,000)
- Few handfuls of garden soil or leaves optional)
- 1 cup of food scraps (fruits, veggies, eggshells, coffee grounds. NO citrus fruit, garlic, or hot peppers)
- 4 bricks

Instructions

1. Drill about 35 holes in the bottom of each plastic tote using a 1/4" drill bit. This is for drainage and for the worms to migrate upward from a full bin to a new, empty bin.
2. Use the 1/16" drill bit to drill about 60 ventilation holes just under the top edge. Also drill about 35 ventilation holes in ONE of the lids.
3. Add half of the shredded office and newspaper, moistening it with the spray bottle. It should feel like a wrung out sponge. This is the bedding.
4. Add the worms, and then add a few handfuls of loose garden soil or leaves if desired—not required.
5. Add food scraps and spread it all out evenly.
6. Top with the other half of the shredded paper and moisten well.
7. Locate your bin's ideal permanent place. The worms don't like temperature extremes, such as really hot summers and really cold winters. Generally, 55 to 75 degrees Fahrenheit (F) is ideal. We keep ours in the attached garage. Dark and or shady environments are also preferred.
8. Place the lid without the holes on the ground and place a brick in each corner.
9. Set the full bin on top of the bricks and close it with the lid (the one with holes).
10. To feed the worms, pull away the shredded paper on top, pour in the cup of food, and then replace the shredded paper so all food is covered.

If fruit flies or mites develop, it is a sign that you may be giving them more food than they can handle. Simply stop feeding them for a few weeks until the pests go away and the conditions balance out.

Over time, the bedding (shredded paper) will break down. Just add more shredded paper if that happens, moistening it with the spray bottle. Monitor the moisture level. The contents should always feel like a wrung-out sponge. If lots of liquid seems to be draining out the bottom, it could mean that you've added too much water. Just let the moisture levels balance out on their own, perhaps lifting the lid once a day to allow excess moisture to escape.

How the Worm Bin Works

Vermicomposting occurs in one main bin. When the bin gets full, stop feeding the worms for about one month. They will continue to process the existing food scraps. By the end of the month, they will be hungry and willing to travel to find a food source. Take the lid off, and place the second empty bin directly on top of the compost surface of the (bottom) full bin.

Add bedding and food (always keep food covered) to the new bin on top, and wait two months for the worms to migrate up through the holes to the new bin. Only add more food if it appears that a good portion of the worms have migrated up. The bottom bin should be mostly worm castings at this point, and easily harvestable for the garden. If you notice a few worms or uncomposted scraps still hanging around in the finished compost, they should be easy to remove and add to the new bin to continue composting. Use the worm compost to improve garden soil or to start seedlings in pots.

Living on a small lot limits my ability to compost in an ideal way, but if I "had my druthers," I would have a compost heap for sticks and brush. The 3-bin compost unit would make amazing compost out of food scraps, leaves, grass clippings, and weeds. And my worm bins (two of them) would produce fertilizer-rich worm castings that, when mixed with finished compost, would make building garden soil a breeze for very little cost and external inputs. All in all, the combination of composting methods you use will be dependent on how much space you have and other unique factors.

Through no-till farming, an informed use of soil amendments, and active composting systems, you'll be well on your way to creating a healthy soil ecology that will form the basis of a thriving food production system.

4 VEGETABLES ON THE MICRO-FARM

THE FIRST SUPERMARKET SUPPOSEDLY APPEARED ON THE AMERICAN LANDSCAPE IN 1946. UNTIL THEN, WHERE WAS ALL THE FOOD? THE FOOD WAS IN HOMES, GARDENS, LOCAL FIELDS, AND FORESTS. IT WAS NEAR KITCHENS, NEAR TABLES, NEAR BEDSIDES. IT WAS IN THE PANTRY, THE CELLAR, THE BACKYARD.

JOEL SALATIN

THE SUBURBAN MICRO-FARM

Small gardens can be surprisingly productive. When I first began gardening in our yard, I didn't believe this was true, and I was discouraged by our lack of space. Not only was our yard extremely small, but we were also graced with a lot of shade from the neighbors' trees. I assumed it would be a waste of time trying to grow anything, so I leased land from an acquaintance that had land she wasn't using. I drove 40 minutes each day to manage my gardens there. In return for the land, I shared my harvest with the property owner. By the end of that summer, I realized how much time I had been wasting in my car: I could have harvested the same yield in my own small yard, even in my partially shady conditions!

In this chapter, we'll look at basic planting suggestions for common vegetables, and some different ways to farm your yard, even with challenging conditions. With an eye toward creativity, there should be something for just about everyone. I'll also share some helpful tips for choosing what to grow, and some techniques that will help you improve your production throughout the season.

The Efficient Micro-Farm

When you concentrate your growing to a compact space, you minimize travel distances for walking and moving materials. To capitalize on efficiency, you can do other things to minimize work, such as keeping livestock and composting operations close to the garden, so that manure and compost can easily be transferred to the garden without as much work. When all of these features are clustered together and reasonably close to the house and garden shed, you can minimize the distance of even more trips for irrigation, tools, bathroom breaks, etc.

Think carefully about all of the materials and supplies you will need to manage each of your farming systems, and figure out how to organize them in clusters. An orchard, for example, which doesn't need nearly as much attention as the vegetable garden, can be farther away from the house, but it should be near the tool shed. An herb garden conveniently located right outside the back door may increase your chances of using fresh herbs in your cooking. A pollination garden near a beekeeping operation would improve the chances of a successful honey harvest without much work on your part.

Observing our micro-farm and searching for connections between the elements is one of the foundations of permaculture. Placing my compost system near the rear property line made the bins easily accessible for my neighbors. At least three neighbors regularly added their yard waste to our bins, which meant more organic matter for my gardens without any additional

work on my part. For more ideas on incorporating permaculture into your micro-farm system, see chapter 9.

A collection of raised beds located near a tool shed and composting area can be the backbone of a very efficient and productive micro-farm.

Can you find ways to intercept useful flows of material for your garden? Some examples would be collecting leaf bags from the curb each fall, collecting coffee grounds from a local coffee shop, collecting pallets from local businesses for micro-farm projects, etc.

Take advantage of the useful flow of water from your roof. My 1,200-square-foot roof collects over 30,000 gallons of rainwater each year. Finding ways to incorporate that water into my garden has allowed me to take advantage of a free resource that reduces watering costs.

Raised beds are helpful in many gardening situations.

Raised Bed Gardening

When planned properly, a few raised beds can keep a household eating fresh produce all summer long. Add a few more, and a family can expect to put away some produce for wintertime eating. The height of the raised bed gives space to improve the soil by adding compost and other amendments. Vegetables will enjoy the loose soil in which to grow deep roots. Raised beds heat up faster in the spring, giving a longer growing season in areas with long winters, and they will drain better during wet seasons. They are also useful for growing on rocky soils, where digging is a challenge, and they are essential for growing food in areas where soil is contaminated.

For beginners, my advice is to start small. Two 3x6-foot raised beds are plenty of space for first-time gardeners. You can always add more garden beds later as your experience and availability permit. To begin, choose a spot in full sun (or the sunniest spot you can find) to place two 3x6-foot raised beds with a 36-inch pathway in between. It is ideal if the long sides of the raised beds face north-south to give your vegetables the most amount of sun exposure.

Raised beds are often made out of wood but can also be made out of concrete blocks, boulders, or anything else you might have. Build them directly over grass or even on top of a paved level surface. The height of raised beds can vary from six inches to as high as you want to make them. Twelve inches is a popular height for raised beds on the ground. For raised beds on top of pavement or in areas with polluted soil, I recommend at least 24-inch-high raised beds to allow space for plant roots. We have two raised beds on our driveway of this height, and they have performed very well.

Filling a Raised Bed on Grass

Use a digging fork to aerate the ground inside the raised bed. Then, cover the grass on the inside with cardboard, overlapping the ends to keep grass and other weeds from growing through. If you are dealing with toxic soil, it will be a good idea to use a weed-blocking landscape fabric, which will prevent your new soil from mixing with the contaminated soil underneath. Now fill each bed, alternating layers of compost soil with layers of organic matter, such as aged manure, grass, straw, shredded leaves, or composted wood chips. The top six inches should be compost soil. This process is called sheet mulching, or lasagna gardening, and it is a cost-effective way to build soil. It is outlined in more detail in chapter 3.

Filling a Raised Bed on Pavement

Cover the bottom six inches of your raised bed with gravel. This will allow good drainage while preventing the soil from washing away. On top of the gravel, follow the instructions above for

adding compost soil and layers of other organic materials. Remember, the taller your raised bed, the more soil and organic matter you will need.

Let your beds sit for at least two weeks and up to three months before planting. For this reason, fall is the best time to build your raised beds. During that time, the soil will settle, so be prepared to add more soil before planting in the spring.

Each fall as you put your garden to bed for the winter, it will be wise to add compost soil, manure, or other organic materials to replace the nutrients lost during the growing season. Cover the soil with mulch to keep the beneficial soil microorganisms sheltered. Alternatively, consider using a cover crop over the winter. In the spring, use a digging fork to loosen the soil and blend in any soil amendments, breaking up clumps.

Sunken Bed Gardening

While raised beds are ideal for areas with long, cold winters and/or rainy springs, sunken beds may be ideal for hot, dry climates. A sunken bed is created by digging down about 18 inches to break up the soil, retaining the topsoil but removing some of the clay and rocks. Add compost soil to amend the growing area. Use the removed clay and rock to build a raised path around the bed. This raised path will protect the sunken bed from drying winds and will keep the bed cooler during hot summers. Moisture and runoff will be encouraged to collect in the sunken beds, reducing irrigation needs.

Raised bed construction on driveway

Permanent Pathways

Raised bed structures and sunken beds create a distinction between the garden beds and the surrounding permanent pathways, offering an organized appearance and helping to keep foot traffic out of the garden beds. Walking or stepping in the beds is strongly discouraged, as it will compact the soil and destroy microorganism habitat.

Sunken Bed

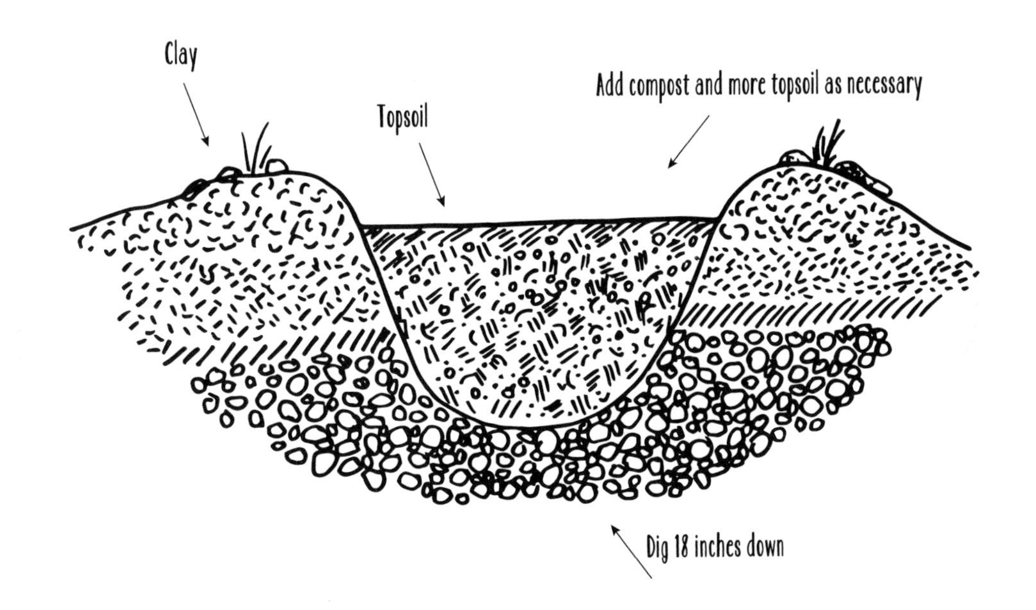

In addition to keeping soil in the garden beds loose, permanent pathways save money on soil amendments. Rather than applying costly amendments over a broad area, with permanent beds you can skip adding amendments to the walkways, only adding them to the permanent garden beds. Permanent pathways of white clover or wood chips attract beneficial insects and fertilize the garden, too. As a general rule, I use a minimum of 36-inch-wide paths to accommodate wheelbarrows and foot traffic in my garden. If you have the space, wider pathways will make the garden more manageable.

Square Foot Gardening

One easy growing method that I've had success with in my raised beds is called square foot gardening, made popular by its founder, Mel Bartholomew. In this method, the bed is divided into square-foot quadrants. Using string or a lattice-type grid can help visualize the quadrants. In a 3x6-foot raised bed, for example, there are 18 (one-foot) squares or blocks in which to plant. In his book *All New Square Foot Gardening*, Bartholomew details how many of each type of vegetable to plant per square foot.

Raised beds using square foot gardening method (Photo by Patrick via Flickr)

Place tall plants (tomatoes, beans, and cucumbers, etc.) on the north side of the bed so they don't shade out plants in the rest of the bed. Place shorter plants on the south side.

Properly locating components—such as a raised bed vegetable garden—is essential to an efficient micro-farm.

 Guide to Planting and Growing Vegetables

In this section, you'll find planting details and instruction for many popular vegetables. Your frost date will help you determine when to plant the crops of your choice. To find your frost date and for more planting details, download the **Seed Starting & Planting Worksheet** at www.TenthAcreFarm.com/tsmf-companion. See chapter 9 for harvesting tips.

Asparagus: A long-lived perennial vegetable that is one of the first harvests of spring. Asparagus is grown by crowns, which are one-year-old plants. Keep in mind that the first harvest will be two years after planting. Choose a location in full sun or partial shade with well-drained, moist soil.

Since asparagus can live for 15 years, you only get one chance to prepare the bed well. Remove all weeds and work compost soil or aged manure into the soil. Dig trenches nine inches wide and five to eight inches deep (shallower in clay soil). Four weeks before your spring frost date, set the crowns

Green beans climb a metal grid attached to the privacy fence.

in the trench on top of a mound, 10 to 15 inches apart. Cover the crowns with two inches of soil. As the spears grow, gradually fill in the trench, keeping three inches of the spears visible.

Beans: A warm season crop that requires well-drained soil and full or partial sun (at least four hours per day). Eggshell powder is a great soil amendment for beans. Mix powdered eggshells and compost soil into the bed two weeks before planting. In the photo at the end of this chapter, hardware cloth is attached to a privacy fence to make a convenient bean trellis on small micro-farms.

Beans come in two types: bush beans and pole beans. Pole beans are twining vegetables that need a trellis, heavy-gauge wire livestock fence panel, or tall stake to climb. Bush beans do not need support. Sow bean seeds one week after your frost date. Plant pole beans one inch deep and three inches apart, bush beans one inch deep and two inches apart, three inches apart if planting in partial shade. Succession planting: Sow beans every 10 days until 10 weeks after your spring frost date.

Broccoli

Beets: A cool-season root vegetable. They require a well-prepared bed with fertile soil in full or partial sun (at least four hours per day). Worm castings and aged manure will improve your beet harvest. Mix the castings and manure into the soil two weeks before planting. Sow seeds outdoors two weeks before your frost date, ½ inch deep, one inch apart (farther apart in partial sun). Succession planting: Sow beet seeds every 14 days until nine weeks before your fall frost date.

Broccoli: A cool-season crop that requires fertile soil and a well-prepared bed in full sun. Eggshell powder and aged manure are both great soil amendments for broccoli. Mix powdered eggshells and manure into the bed two weeks before planting.

Start seeds indoors nine weeks before your frost date. Transplant seedlings or sow broccoli seeds outdoors, ½-inch deep and 18 inches apart, two weeks before your spring frost date. For a fall harvest, start broccoli seeds indoors six weeks after your spring frost date. Transplant seedlings or sow broccoli seeds outdoors 11 weeks before your fall frost date.

Brussels sprouts

Brussels sprouts: An excellent fall crop harvested after a frost with a long growing season. They can be started indoors or sown directly outdoors. Brussels sprouts prefer fertile, moist soil in full or partial sun

(at least four hours per day). Start seeds indoors six weeks before your frost date. Transplant seedlings or sow seeds outdoors, ¼-inch deep and 18 inches apart (farther apart in partial sun), two weeks before your spring frost date.

For a fall harvest, start seeds indoors five weeks after your spring frost date. Transplant seedlings or sow seeds outdoors nine weeks before your fall frost date.

A bed of maturing carrots

Cabbage: A cool season crop requiring fertile, well-prepared soil in full or partial sun (at least four hours per day). Mix aged manure into the soil two weeks before planting.

It can be started indoors or sown directly outdoors. Start seeds indoors six weeks before your frost date. Transplant seedlings two weeks before your frost date, or sow seeds outdoors the week of your frost date, ¼-inch deep and 18 inches apart (farther apart in partial sun). For a fall harvest, start seeds indoors five weeks after your spring frost date. Transplant seedlings or sow seeds outdoors nine weeks before your fall frost date.

Cabbage

Carrots: A cool season root crop that prefers fertile, loose soil in a well-drained bed with full or partial sun (at least four hours per day). Mix worm castings and compost into the soil two weeks before planting. Sow outdoors two weeks before your frost date, ¼-inch deep and three inches apart (four inches in partial sun). For a fall harvest, sow seeds outdoors 11 weeks before your fall frost date. The photo at the end of this chapter shows carrots growing with calendula.

Cauliflower: A cool season crop that requires a long season and cool temperatures to produce. It prefers rich, fertile, well-drained soil in full or partial sun (at least four hours per day). Worm castings, aged manure, and powdered eggshells are all great soil amendments for cauliflower. Mix them into the soil two weeks before planting.

Start seeds indoors six weeks before your frost date. Transplant seedlings two weeks before your frost date, 18 inches apart (farther apart in partial sun). For a fall harvest, start seeds indoors six weeks after your spring frost date. Transplant seedlings 11 weeks before your fall frost date.

Cauliflower

Celery: A cool season crop that requires a long season and consistent cool temperatures to produce. It requires fertile, moist soil in full or

Collard greens

partial sun (at least four hours per day). Mix compost into the soil two weeks before planting.

Start seeds indoors 10 weeks before your frost date. Transplant seedlings four weeks after your frost date, 10 inches apart (12 inches in partial sun). For a fall harvest, start seeds indoors three weeks before your spring frost date. Transplant seedlings 11 weeks after your spring frost date.

Chard: Requires well-drained, fertile soil in full or partial sun (at least four hours per day). Seeds can be started indoors or sown outdoors. Start seeds indoors four weeks before your spring frost date. Transplant seedlings or sow seeds outside, ½-inch deep and 14 inches apart, the week of your spring frost date. For a fall harvest, sow seeds nine weeks before your fall frost date.

Corn

Collard greens: A cool season crop that requires rich and fertile, well-drained soil. Collards will grow well in full or partial sun (at least four hours per day). Seeds can be started indoors or out. Start seeds indoors six weeks before your frost date, and transplant them outside two weeks before your frost date. Collard greens will grow right through winter using season extension techniques like cold frames and row covers.

Or start seeds outdoors by sowing them ¼-inch deep and 12 inches apart, the week of your spring frost date. Succession planting: Direct sow collard greens outdoors every three weeks until 11 weeks before your fall frost date. For a fall harvest, start seeds indoors 10 weeks after your spring frost date. Transplant seedlings or direct sow seeds outdoors 11 weeks before your fall frost date.

Cucumbers

Corn: A warm season crop that requires a long growing season, full sun, and rich soil. Worm castings and aged manure will benefit a corn crop. Mix them into the soil two weeks before planting. Corn requires good pollination to produce ears, so plant corn in blocks rather than long rows. Sow seeds outdoors one inch deep and 10 inches apart, two weeks after the frost date. Succession planting: Corn can be sown every 10 days until nine weeks after the spring frost date.

Cucumbers: A warm season crop that requires full sun and rich, well-drained soil. Worm castings, aged manure, and powdered eggshells are all great soil amendments for cucumbers. Mix them into the soil two weeks before planting.

Cucumbers can be started indoors or out. Start seeds indoors three weeks before the spring frost date. Transplant seedlings or sow seeds outdoors, ½-inch deep and 12 inches apart, two weeks after the frost date. Succession planting: Sow or plant cucumbers every three weeks until eight weeks after your spring frost date.

Eggplant

Eggplants: A warm season crop that prefers full sun, a well-prepared bed, and well-drained soil. Start seeds indoors eight weeks before your frost date. Transplant seedlings, 24 inches apart, four weeks after your frost date.

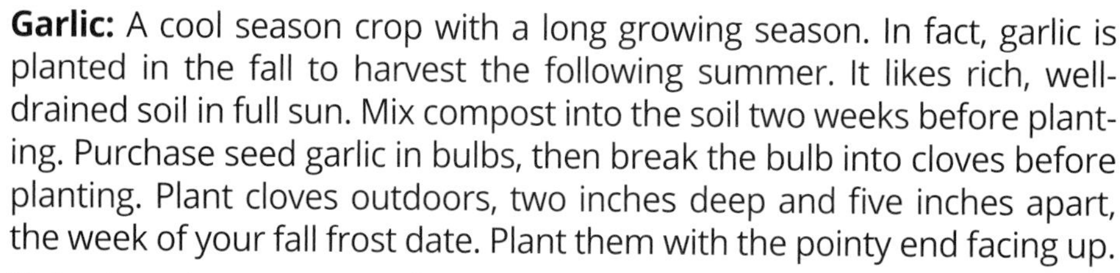

Garlic: A cool season crop with a long growing season. In fact, garlic is planted in the fall to harvest the following summer. It likes rich, well-drained soil in full sun. Mix compost into the soil two weeks before planting. Purchase seed garlic in bulbs, then break the bulb into cloves before planting. Plant cloves outdoors, two inches deep and five inches apart, the week of your fall frost date. Plant them with the pointy end facing up.

Garlic

Kale: A cool season crop that requires rich and fertile, well-drained soil. Kale will grow well in full or partial sun (at least four hours per day). Seeds can be started indoors or out. Start seeds indoors nine weeks before your frost date. Transplant seedlings or sow seeds outdoors, ¼-inch deep and 12 inches apart, four weeks before your frost date.

Succession planting: Direct sow kale seeds outdoors every three weeks until 11 weeks before your fall frost date. For a fall harvest, start seeds indoors seven weeks after your spring frost date. Transplant seedlings or direct sow seeds outdoors 11 weeks before your fall frost date.

Kohlrabi: A cool season crop that requires rich, moist soil and full or partial sun (at least four hours per day). Mix compost into the soil two weeks before planting. Sow seeds outdoors, ¼-inch deep and eight inches apart, two weeks before your frost date. Succession planting: Sow kohlrabi every 10 days until eight weeks before your fall frost date.

Kale

Leeks: A cool season crop that needs a long season in fertile, moist soil and full sun. Mix aged manure into the soil two weeks before planting. Seeds can be started indoors or transplant purchased starts outdoors. Start seeds indoors 10 weeks before your frost date. Transplant seedlings/starts outdoors, six inches deep and six inches apart, the week of your frost date. For a winter harvest, start seeds indoors the week

Lettuce

of your frost date and transplant them outdoors 10 weeks after your spring frost date.

Lettuce: A cool season crop that prefers loose, fertile, moist soil and full or partial sun (at least four hours per day). Worm castings, compost, and powdered eggshells are all great soil amendments for lettuce. Mix them into the soil two weeks before planting.

Lettuce can be started indoors or out. Start seeds indoors six weeks before your frost date. Transplant seedlings or sow seeds outdoors, on top of soil surface and 10 inches apart, four weeks before frost date. Press seeds in well. Succession planting: For a continual lettuce harvest, start or sow lettuce seeds every 10 days until nine weeks before fall frost date.

Melons

Melons: (cantaloupe, watermelon) A warm season crop that requires fertile, well-drained soil in full sun. Mix aged manure into the soil two weeks before planting. Direct sow seeds outdoors, ½-inch deep and 36 inches apart, four weeks after your frost date. Succession planting: Sow melon seeds every three weeks until 10 weeks after your spring frost date.

Okra: A warm season crop that prefers fertile, well-drained soil in full sun. Mix aged manure into the soil two weeks before planting. Start seeds indoors two weeks before your frost date and transplant them outside, 18 inches apart, four weeks after your frost date. Start seeds indoors as late as four weeks after the frost date, transplanting them outdoors as late as 10 weeks after the spring frost date.

Okra

Onions: A cool season crop that prefers rich, well-drained soil in full sun. Onions will enjoy aged manure and worm castings. Mix them both into the soil two weeks before planting. Onion varieties must be chosen for their appropriate day length. Short day onions are appropriate for southern gardeners (zone 7 and warmer), day-neutral or intermediate varieties will work for those in middle regions (zones 5 & 6), while long day onions are appropriate for northern gardeners (zone 6 and colder).

In addition, onions can be planted from seed, as transplants, or as sets. I've had the best luck with onion transplants, which are onion

seedlings that were grown in the early spring and sold to gardeners to plant outside two to four weeks before their spring frost date.

Different gardeners swear by different methods, though. Some swear that starting your own onions by seed indoors create the biggest bulbs, while others claim onion sets are the way to go. Sets are immature young onion plants grown the previous year, forced into dormancy, and sold to gardeners in the spring to plant at the same time as transplants. You'll have to experiment for yourself to figure out whether seeds, sets or transplants produce the best onions for you.

Onions

Start onion seeds indoors 10 weeks before your frost date. Transplant your own seedlings, purchased sets, or transplants, one inch deep and five inches apart, outdoors two to four weeks before your spring frost date. Transplants and sets can be planted until eight weeks after your spring frost date.

Peas: A cool season crop that prefers fertile, well-drained soil and full to partial sun (at least four hours of sun per day). Peas will enjoy aged manure and worm castings. Mix them both into the soil two weeks before planting. There are three types of peas: shelling (in which the pod is discarded prior to eating), snap (in which the pod is eaten), and snow peas (in which there is more pod than pea).

Pea vines

Pea vines are climbers, but they do not need support unless they grow taller than three feet. All peas are planted in the same way. Sow pea seeds outdoors, one inch deep and two inches apart, four weeks before your frost date. Succession planting: sow seeds every 14 days until 12 weeks after your spring frost date to get the best harvest. In the photo at the end of this chapter, the peas are climbing an arbor leading to the front porch.

Peppers: A warm season crop that requires well-drained, fertile soil in full sun. Mix worm castings and compost into the soil two weeks before planting. Start seeds indoors eight weeks before your frost date. Transplant seedlings outside, 18 inches apart, two weeks after your frost date. Start seeds indoors as late as the week of your spring frost date, transplanting them outdoors as late as 10 weeks after your spring frost date.

Peppers

Potatoes

Potatoes: A cool season crop that prefers well-drained, fertile soil in full or partial sun (at least four hours of sun per day). Mix worm castings and aged manure into the soil two weeks before planting. Potatoes are planted from seed potatoes that are cut into one- to two-inch pieces with at least two eyes per piece. These potato pieces are cut one or two days ahead of the planting date. Plant "seeds" eye side up, three inches deep and 12 inches part, two weeks after your frost date.

As the plants reach eight inches tall, mound up the soil over the plants so that only two inches of green stem is showing. Continue this process, called hilling, until the hills are around 12 inches high. This protects the potatoes from the sun. To plant fall potatoes, follow the same procedure, planting seeds eight weeks after your spring frost date.

Radishes

Radish: A fast-maturing, cool season root vegetable that prefers fertile, loose, well-drained soil and sun to partial sun (at least four hours of sun per day). Mix aged manure into the soil two weeks before planting. Sow radishes outside, ½-inch deep and two inches apart, two weeks before your spring frost date. Succession planting: For a continuous harvest, sow seeds every 14 days until five weeks before your fall frost date.

Rhubarb: A perennial vegetable producing popular stalks with a tart flavor. Rhubarb prefers fertile, well-drained soil and full sun. Mix aged manure into the soil two weeks before planting. Rhubarb is planted from crowns, which are one-year-old roots. Remove all weeds from the planting site and work in compost or aged manure prior to planting. Crowns are planted two inches deep and four feet apart, any time between four weeks before your spring frost date and three weeks after your fall frost date.

Rhubarb

Rutabaga: A cool season vegetable that requires well-drained, fertile soil and full sun. Sow seeds outdoors, 1/3-inch deep and six inches apart, three weeks after the frost date. For a fall harvest, sow seeds 12 weeks before your fall frost date.

Spinach: A cool season vegetable that requires fertile, moist soil and full or partial sun (at least four hours of sun per day). Mix in aged manure two weeks before planting. Sow seeds outdoors, ½-inch deep and five inches apart, two weeks before your spring frost date. Succession

planting: Sow seeds every 14 days. For a fall harvest, sow seeds seven weeks before your fall frost date.

Summer squash: (yellow squash, zucchini) A warm season crop that prefers rich, well-drained soil in full sun. Mix worm castings and compost into the soil two weeks before planting. Summer squashes are vining plants that take up a lot of room. Sow seeds outdoors, one inch deep and 24 inches apart, four weeks after your frost date.

Sow seeds again 10 weeks after your spring frost date. This second sowing ensures that some plants will survive in case of pest problems. There are bush varieties of many summer squashes that take up less space in the garden.

Ripening heirloom tomatoes

Sweet potatoes: A warm season crop with a long growing season that prefers well-drained, loose soil in full sun. Mix worm castings and compost soil into the soil two weeks before planting. Sweet potatoes are planted from bare roots called slips. Plant slips deep enough so that the roots and about ½ inch of the stem is buried, about three to four inches deep in total.

Slips should be spaced about 18 inches apart. Sweet potato plants can be trained to grow up a trellis to save space. If growing on the ground, frequently lift up the vines to discourage them from rooting. The latest safe planting date for sweet potato slips is 10 weeks after your spring frost date.

Turnips

Tomatoes: A warm season crop that prefers well-drained, rich soil with full sun. Mix worm castings and aged manure into the soil two weeks before planting. Start seeds indoors six weeks before your frost date. Transplant seedlings outdoors two weeks after your frost date, spaced 20 inches apart. Tomato seedlings can be planted as late as 10 weeks after your frost date.

Turnips: A cool season root crop that requires loose, well-drained, rich soil in full or partial sun (at least four hours per day). Turnips will enjoy compost and worm castings. Mix both into the soil two weeks before planting. Sow seeds outside, ½-inch deep and eight inches apart, two weeks before your spring frost date. For a fall harvest, sow seeds again eight weeks before your fall frost date.

Zucchini

Winter squash: (acorn, butternut, pumpkin) A warm season crop with a long growing season that prefers rich, well-drained soil in full sun. Mix worm castings and aged manure into the soil two weeks before planting.

Winter squashes are vining plants that take up a lot of room. Sow seeds outdoors, one inch deep and 36 inches apart, four weeks after your frost date. Sow seeds again 10 weeks after your spring frost date. This second sowing ensures that some plants will survive in case of pest problems.

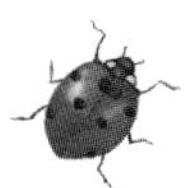

Choosing Crops to Meet Your Needs

If you are dealing with shade, garden pests, small spaces, or no space at all, it can be frustrating to read about building endless raised beds in a bright and sunny location and planting as much as the heart desires. The "Container Crops" table shows just a few of the ways you can take advantage of those less than optimal conditions and make the best decision on how much to plant.

Container Crops		
Container Crops for Sun	Container Crops for Partial Sun/Shade	Container Crops for Shade
eggplant okra peppers tomatoes potatoes	basil beets broccoli Brussels sprouts leafy greens (chard, collards, kale, spinach) parsley peas pole beans radishes salad greens strawberries	celery onions

Edible Container Gardening

If you're lacking garden space, growing edibles in containers may be a solution. Container gardening is becoming easier as more and more container-friendly varieties are being discovered.

CONTAINER GARDENING TIPS

- Use the largest containers possible and good potting soil.
- Use liquid fertilizer once a week starting midsummer, unless the potting soil already contains fertilizer.

Growing Vertically

You can save space in a small garden by growing vertically. Climbing and vining vegetables grown on a trellis or other vertical structure can produce higher yields with a reduced susceptibility to plant disease. They are also beautiful and provide architectural dimension to the garden.

VEGETABLES TO GROW VERTICALLY

- cantaloupe
- cucumbers
- green beans
- peas
- squash (acorn and butternut)
- sweet potatoes (some varieties)
- tomatoes
- watermelon

Choose the Right Trellis for Climbing Vegetables

Did you know that different vines climb or spread in different ways? I used to assume all trellising structures were created equal, until I noticed that certain climbing vegetables just didn't seem to take to the structure I was providing for it.

Winter squash growing on A-frame trellis made with twine

Vegetables with Tendrils

Cucumbers, peas, and many vining squash varieties have tendrils that reach out from the plant's stem in search of something to grab onto and climb. The tendrils can go upwards and sideways. Tendrils prefer to grab onto something organic and non-metallic, such as twine (their favorite) or wood lattice (try hand-making one from tree branches).

I like to use twine to create a grid-like structure between two posts for them to climb. Grid squares should be less than four inches in width/diameter. You can make a metal trellis friendly to the tendril climber by wrapping twine around the metal supports and creating a twine grid in open spaces. A teepee or A-frame trellis made with stakes and twine is sturdy enough to support the heavier weight of winter squash.

Beans growing on a bamboo teepee

Vegetables that Twine

Pole beans are twiners, meaning that as the vining stem grows upward, it will wrap itself around anything it can touch. Twiners are not picky about what they climb, so you can grow them on any type of trellis, support structure, or fence that you want. Just be sure it is tall, as they can reach almost indefinitely. Many gardeners pinch off the ends of the vines when they reach the top of the trellis.

One interesting thing to note is that pole beans twine in a counterclockwise habit. If you are training it to grow up a certain support, observe your vine's pattern and be sure to twine it in the direction that would be natural for the vine.

Vegetables that Scramble

Tomatoes and sweet potatoes are not in the same family, but they are both non-climbing, scrambling vines. If left to their own devices, they'll scramble happily along the ground, rooting in the ground at nodes along the vine. Supporting tomatoes with trellises or cages is common. Training the plant to grow upright means that it will put more energy into making tomatoes than it will into rooting itself like it does when it scrambles. That's good news for us!

Try growing sweet potatoes vertically to save space, too. They'll have to be trained, since—like tomatoes—they do not have tendrils nor do they twine. I like to use stake-and-twine teepees or A-frame structures to provide sturdy support, so I can weave the vines in and out of the twine grid as they grow. This will produce larger tubers, since the plants will put more energy into the existing tubers, rather than growing a bunch of little ones.

Front yard sweet potato teepee

Shade Tolerant Micro-Farming

Many suburbanites deal with shady conditions in their leafy neighborhoods. I much prefer tree-filled landscapes to the lawn environment, but they certainly can throw a wrench in our farming pursuits.

At the beginning of this chapter, I mentioned my struggle to balance my micro-farming endeavors with the realities of our tree-covered backyard. After some trial and error, and some accidental successes, I soon discovered the opportunities of shade. In just one year, I grew 80 pounds of leafy greens and root vegetables in two raised beds under shady conditions.

SHADE TOLERANT CROPS				
Leaf Crops	**Root Crops**	**Stem & Shoot Crops**	**Legumes**	**Herbs**
cabbage chard collards kale lettuce spinach	beets carrots kohlrabi potatoes radishes turnips	Brussels sprouts cauliflower celery	beans peas	basil chives cilantro lemon balm marjoram mint oregano parsley

There are also many food-producing perennials that can tolerate partial shade. Our black raspberries and currant bushes grow in mostly shaded areas. Strawberries and asparagus grow in dappled shade. Six hours of sunlight is the minimum ideal for fruit trees, but tart cherry trees and pawpaws may do okay with less than that. Check nursery catalogs to find out if these choices are well-suited for your growing zone.

If your yard is heavily shaded, you may want to consider adding some edible forest plants. Mushrooms such as shiitakes and wine caps are easy varieties for beginners to try. Ramps will do well in an edible forest garden, too. Both mushrooms and ramps would go for a high price to local chefs if you were so inclined. Though you might get a smaller harvest, shade-tolerant plants will produce a crop in areas that receive at least 3 to 4 hours of sunlight per day. The "Shade Tolerant Crops" table lists some shade-tolerant vegetables and herbs.

Planting with Deer and Rabbits

Are you struggling to plant a garden but continue to be thwarted by these furry pests? While good fencing may be the only sure-fire way to keep these persistent creatures out of your garden, here are some suggestions for what to plant—and not to plant—when dealing with these animals. Resistant vegetables are not animal proof, but they have a better chance of withstanding animal pressure than others. The herbs—when sprinkled around the garden—can help to protect your resistant vegetables. Avoid the rabbit and deer favorites if you struggle to garden in their presence.

Rabbit-Resistant Veggies	Rabbit-Resistant Herbs	Rabbit Favorites (Avoid)
asparagus leeks onions potatoes squash tomatoes	anise hyssop bee balm chives garlic lavender yarrow	leafy greens (chard, collards, kale, lettuce, spinach) strawberries

Deer-Resistant Veggies	Deer-Resistant Herbs	Deer Favorites (Avoid)
asparagus carrots cucumbers eggplant fennel garlic leeks onion peppers rhubarb tomatoes	chives dill lavender lemon balm mint parsley rosemary sage thyme	beans fruit (all kinds) leafy greens (chard, collards, kale, lettuce, spinach) peas

How Much to Plant?

How many of each variety, how many different types of vegetables to plant—this depends on your commitment to self-sufficiency, your dietary habits, and your time availability. Do you want to grow ALL your own produce for the entire year? Are you a picky eater? John Jeavons, in *How to Grow More Vegetables*, approximates that you need 4,000 square feet of growing space to grow all of one person's produce for an entire year.

Even if you don't have much space, you can choose some of your favorites to grow. Still, you might consistently over-plant certain crops and feel overwhelmed at harvest time. The "How Much to Plant" table is a rough guide on how much to plant PER PERSON to provide a full year's supply.

This gives you some idea of how much yield you will get from different crops—two melon plants are all you need to harvest enough melons to satisfy the average melon lover, which is achievable in a decent-sized, sunny backyard. But 80 onion plants might not be the most effective use of your limited suburban space. Deciding what to plant will always involve tradeoffs, and you'll need to plan based on what makes the most sense for your circumstances.

HOW MUCH TO PLANT			
Vegetable	**Quantity**	**Vegetable**	**Quantity**
asparagus	15	kohlrabi	5
beans (bush)	15	leeks	15
beans (pole)	10	lettuce	12
beets	36	melons	2
broccoli	10	okra	3
Brussels sprouts	8	onions	80
cabbage	10	peas	60
carrots	100	peppers	5
cauliflower	5	potatoes	30
celery	8	pumpkins	1
chard	4	radish	20
collards	5	rhubarb	2
corn	40	rutabaga	10
cucumbers	3	spinach	20
eggplant	3	squash	2
garlic	16	sweet potatoes	10
kale	4	tomatoes	5
		turnips	10

Choosing Crops by Hardiness

When deciding what vegetables to plant when, think of them as being cold-loving or heat-loving. Vegetables are often categorized by their hardiness to frost, which helps determine whether to plant them in spring/fall/winter or in summer after all danger of frost has past. Cold-loving crops can be either *hardy* (withstanding frosts down to 25 degrees Fahrenheit) or *semi-hardy* (withstanding frosts down to 29 degrees). Hardy vegetables are well-suited for spring/fall/winter gardens, while semi-hardy vegetables are ideal for spring/fall gardens. Heat-loving vegetables—called *tender*—are sensitive to frost and enjoy the heat of summer.

Hardy Vegetables	Semi-Hardy Vegetables	Tender Vegetables
broccoli Brussels sprouts cabbage collards kale kohlrabi leeks parsley peas radish spinach turnip	beets carrot cauliflower celery chard lettuce potatoes rutabaga	beans corn cucumber eggplant melons okra peppers pumpkins squash sweet potatoes tomatoes

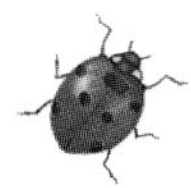

How to Place Vegetables

To ensure the biggest harvest for the time you spend in the garden, practices such as companion planting, succession planting, and crop rotation will really help. Companion planting will help increase yields by using certain plant combinations. Succession planting will increase yields by properly timing the sowing of seeds, and crop rotation will ensure that pests and diseases don't build up in the soil.

Onions are interplanted among collards, beets, and kale to ward off pests.

Companion Planting

Companion planting is an old-time technique for planting a successful garden. Certain plants are planted together which are said to offer specific benefits, such as deterring pests, attracting beneficial insects, or providing a symbiotic relationship at the root level for more vigorous growth. Some common pairings border on folklore, as their benefits haven't been substantiated by science. Yet using this technique—popularized in modern times by Louise Riotte, author of *Carrots Love Tomatoes: Secrets of Companion Planting for Successful Gardening*—gives us a method for organizing our garden and deciding what to plant, where.

In addition to beneficial companions, there are antagonistic plant pairings known to have harmful effects on one another, such as stunting growth or attracting pests. The Companion Planting table shows some of the most popular companions and antagonists.

COMPANION PLANTING			
Vegetable/Family	Vegetable/Fruit Companion	Herb Companion	Antagonists
asparagus	tomato	basil, parsley	onion family
beets	cabbage family, lettuce, onions		beans
cabbage family (broccoli, Brussels sprouts, cabbage, cauliflower, collards, kale, kohlrabi, radish, rutabaga, turnip)	beets, bush beans, carrot, celery, chard, cucumber, lettuce, onion family, potato, spinach	chamomile, dill, nasturtium, oregano	legumes (beans, peas), sage, strawberry, tomato
carrot	onion family	rosemary, sage	
celery	beans, cabbage family, onion family, tomatoes		
corn	cucumber, legumes (beans, peas), melon, potato, pumpkin, squash	sunflower	
cucumber	cabbage family, corn, eggplant, legumes (beans, peas), lettuce, radish, tomato	nasturtium, sunflower	strong-scented herbs, potato
eggplant	legumes, pepper, potato		fennel
legumes (beans, peas)	carrots, cabbage family	calendula, summer savory	
lettuce	carrots, cucumbers, onion family, radishes, strawberries	dill, sunflower	

COMPANION PLANTING			
Vegetable/Family	Vegetable/Fruit Companion	Herb Companion	Antagonists
melons (cantaloupe, watermelon, etc.)	corn	sunflower	potato
okra	eggplant, bell peppers	basil	
onion family	beets, cabbage family, carrot, lettuce, pepper, spinach, strawberry, tomato	chamomile	asparagus, beans, peas, sage
peppers	carrot, onion	basil	fennel, kohlrabi
potato	cabbage family, corn, legumes (beans, peas)	calendula	cucumber, pumpkin, raspberry, rutabaga, squash family, sunflower, tomato, turnip
spinach	cabbage family, celery, legumes (beans, peas), lettuce, onion, radish, strawberry		potato
squash family (yellow squash, zucchini, winter squash, etc.)	beans, celery, corn, onion, radish	nasturtiums	potato
sweet potato	beans, beets, parsnips	dill, oregano, summer savory, thyme	squash
Swiss chard	cabbage family, legumes		
tomato	asparagus, bush beans, carrot, celery, cucumber, head lettuce, onion	basil, calendula, nasturtium, parsley	cabbage family, corn, dill, fennel, pole bean, potato

Succession Planting

Succession planting is a way to sow vegetable seeds to get a continuous harvest all year long. This technique is most often used for cool season vegetables planted by seed. Heat-loving vegetables, such as pepper or tomato seedlings, are usually planted all at once in a small garden in areas with hard winters. The following is an example of succession planting beets:

Beet seeds can be sown in succession every 14 days. Reserve space for four rows of beets in a bed. The first time you sow beets, only sow one row. Then, 14 days later, sow the second row; 14 days after that sow the third row, and so on. This allows you to harvest beets continually rather than all at once.

TIP: SUCCESSION PLANTING

Find more succession planting details on the Seed Starting and Planting Worksheet at www.TenthAcreFarm.com/tsmf-companion.

When you harvest a row of beets, blend in some compost soil, aged manure, or worm castings, and then sow that first row again. I use my monthly calendar to record when I've sown a row so that I can calculate the date of my next sowing action. (Download it at www.TenthAcreFarm.com/tsmf-companion.) Having the written record helps me remember when to harvest and prioritize other tasks in the garden. (Of course, if you'd rather have one big harvest of beets, that's okay, too. Just sow all four rows at the same time!)

The crops shown in the Succession Planting table are sown in spring and can continue to be sown all summer according to their succession planting schedule. Keep in mind that the germination rate will be lower in the hot, dry weather. Try covering seeds with summer weight row fabric and watering daily for improved germination.

While succession planting is most often used for continuously planting vegetable seeds, it can also work for some vegetables that are planted as seedlings. For example, spring-planted broccoli is usually harvested in July (in my region). Then, in early August (or 11 weeks before your first fall frost date) the plants are replaced with new broccoli seedlings for a late fall harvest.

Succession Planting

Beet Profile

Sow seeds 6 inches apart

Sow beets in succession every 14 days

Harvest beets after 40 days

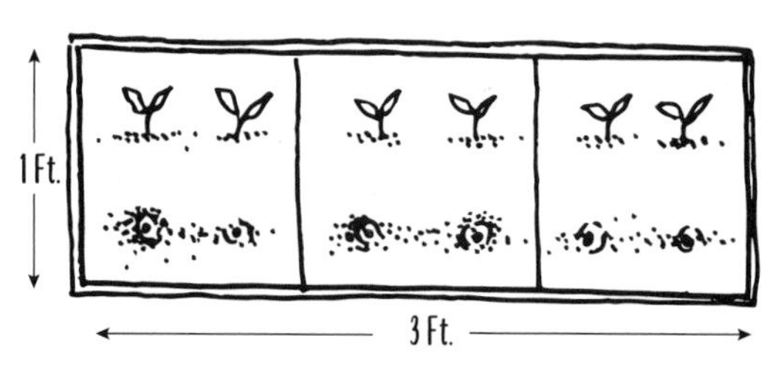

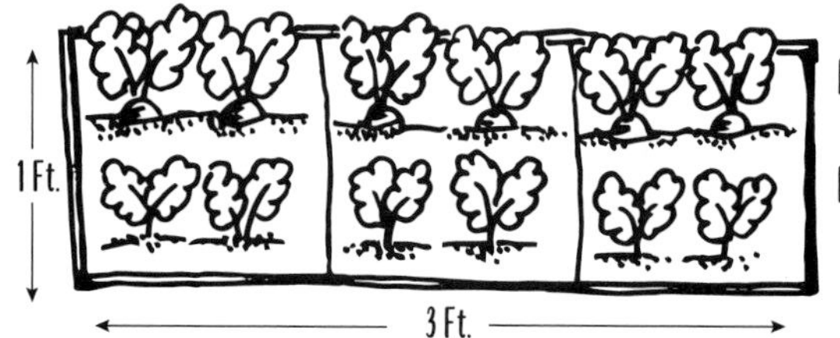

Succession Planting

Spinach Profile

Sow seeds 5 inches apart

Sow spinach in succession every 14 days

Harvest spinach after 35 days

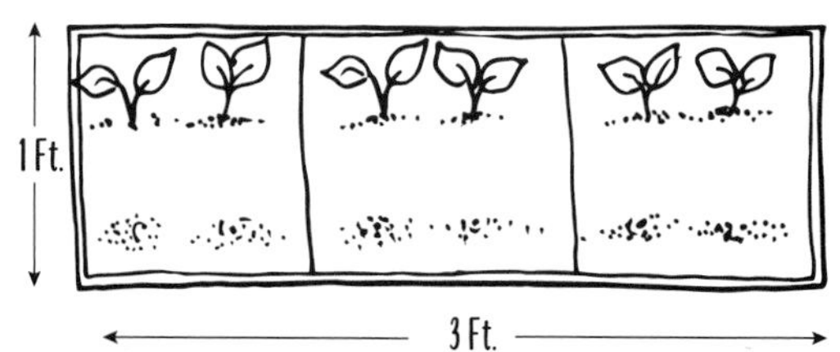

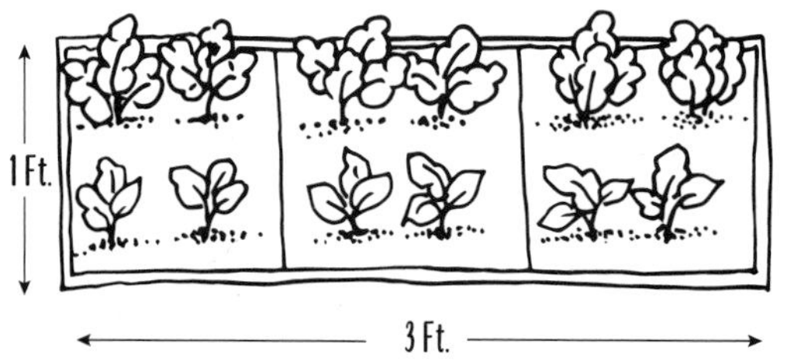

SUCCESSION PLANTING	
Crop	Succession Days (# of days between sowings)
beans	10
beets	14
collards	21
corn	10
cucumber	21
kale	21
kohlrabi	10
lettuce	10
melons	21
peas	14
radish	14
spinach	14

Crop Rotation

If you've planted a garden in the past, you'll want to practice crop rotation to minimize disease and manage soil fertility. Some crops are heavy feeders and will quickly deplete the soil and invite pests if they're planted in the same place each year. Crop rotation is an approach that uses plant families to decide what to plant where in the garden. Keeping good garden records will help you remember and plan accordingly.

Plan to leave at least two to three years between planting members of the same crop family in a particular area of the garden. For example, after planting tomatoes in a particular corner of the garden, I'll wait two years before I plant them in the same place again.

This strategy can be challenging in a small garden or if you have shade and therefore do not have full range of the garden in which to rotate crops each year. In this case, you may not be able to rotate crops appropriately. You can try growing your favorite crop in the same place each year, but if it is overcome by a pest, it will be wise to leave that crop out of the rotation of that bed for at least two years. It's a sign that the soil has been depleted of the nutrients that that plant

uses most, and it needs a rest. [Hint: A couple feet away from the original planting area is not far enough for rotation!] Plant a cover crop and allow that section of garden to rest for a season.

The "Common Crop Families Nutrient Requirements and Rotation Suggestions" table lists the common crop families, their nutrient requirements, and suggestions for what plant families to follow them with.

COMMON CROP FAMILIES, NUTRIENT REQUIREMENTS, AND ROTATION SUGGESTIONS		
Plant Family	Requirements	Follow with These in Rotation
Beet Family: beet chard spinach	Heavy feeders; work in compost or aged manure before planting	Leaf
Cabbage Family: broccoli Brussels sprouts cabbage cauliflower kale radish rutabaga turnip	Heavy feeders; work in compost or aged manure before planting	Legume
Carrot family carrot celery fennel parsley	Light to medium feeders, but do best when compost or aged manure is worked into soil before planting	Legume or Leaf
Squash family cucumber melon squash	Heavy feeders; work in compost or aged manure before planting	Legume
Leaf crops lettuce corn	Heavy feeders; work in compost or aged manure before planting	Any

COMMON CROP FAMILIES, NUTRIENT REQUIREMENTS, AND ROTATION SUGGESTIONS		
Plant Family	Requirements	Follow with These in Rotation
Legume family beans peas clover	Legumes enrich and build soil	Any except onion
Onion family onion garlic leek	Heavy feeders; work in compost or aged manure before planting	Leaf or Cabbage
Nightshade family potato pepper tomato eggplant	Nightshades are heavy feeders; add compost or aged manure before planting nightshades	Carrot, Onion, or Legume

Extending the Season with Cold Frames

Cold frames are small structures that protect cold-loving plants from extreme cold temperatures, like unheated mini greenhouses. A cold frame can warm the soil and air by seven to 10 degrees Fahrenheit, allowing you to easily extend the growing season by four weeks or more. Many people can push this extension even longer, depending on the plant and your growing zone. Cold frames allow you to get a jump on the season in the spring by sowing cold-hardy crops sooner while also extending the harvest in fall and winter.

Making a cold frame can be done inexpensively from many types of common materials or they can be purchased. The common design includes a front face made of transparent, sun-permeable material that is angled south toward the sun for maximum sun exposure and minimal shadows on the interior.

The front face of a cold frame is angled toward the sun to minimize shadows.

Plastic will help to insulate overwintering cold-hardy crops.

Start Cool-Season Crops in the Early Spring

Cold-hardy vegetable seeds can be sown four weeks earlier than normal when using a cold frame. The key to jumpstarting spring with cold frames is to group like crops together in each garden bed: Separate cool-season crops from heat-loving ones and directly sown seeds from the seeds you will start indoors.

The cold frame will be the permanent home of the seeds you sow inside, so choose the crops and cold frame placement wisely. For cool-season vegetables, close the cold frame when the temperature dips below 32 degrees, otherwise, keep the lid cracked for air movement. Cool-season vegetables in a cold frame will be able to survive on their own starting at about two weeks *before* your spring frost date. Disassemble your cold frame at this time and store it until fall.

Kale in cold frame

VEGETABLES THAT CAN BE DIRECT-SOWN IN A COLD FRAME

- beets
- broccoli
- Brussels sprouts
- carrots
- kale
- kohlrabi
- lettuce
- peas

Transplanting Seedlings into a Cold Frame

Cool-season seedlings started indoors in the early spring can benefit from using cold frames. Simply start your plants indoors under grow lights four weeks earlier than you normally would, and when you're ready to plant them, move them to the cold frame to be hardened off. Hardening off is the one-week process by which seedlings started indoors are acclimated to the outdoor elements such as temperature fluctuations, wind, direct sunlight, and cold nights.

Use a thermometer to monitor the temperature inside. Move the seedlings in their trays or pots to a closed cold frame for two days. For the next four days, crack the opening during the day if the inner temperature is in the 60s or above, but close the lid at night. After a week inside the cold frame, they're ready to be planted.

Plant the seedlings inside the cold frame. This will be their permanent bed for the season, so again, choose your plants and cold-frame placement wisely. Keep the cold frame vented most of the time, closing it when temperatures dip below freezing. Two weeks *before* your spring frost date, your cool-season vegetable seedlings will be able to survive on their own outdoors. Disassemble the cold frame at this time and store it until fall.

Heat-loving vegetables like eggplant, okra, peppers, and tomatoes can also be started indoors three weeks earlier thanks to a cold frame.

On the same date that you start your seeds indoors, warm up the garden bed where you will permanently plant the seedlings by covering it with a cold frame. Keep the lid closed and lay an inner protective layer on the bed, such as a row cover, blanket, or chemical-free straw. A *cloche* is another implement that can be used to give your plants more protection. *Cloches* are the French term for mini greenhouses that are made to cover individual plants. You can make your own cloches cheaply and easily using milk jugs with the bottoms cut off and set over individual plants.

COOL-SEASON VEGETABLES THAT CAN BE PLANTED AS SEEDLINGS IN A COLD FRAME

- broccoli
- Brussels sprouts
- cabbage
- collards
- kale

Set a thermometer in a shady spot inside the cold frame and monitor the temperature: When it maintains 70 degrees F inside, it's time to plant your seedlings. Follow the hardening-off procedure listed above before transplanting the seedlings into the cold-frame bed.

The temperature inside the cold frame will affect the health of heat-loving seedlings. The ideal temperature for heat-hardy vegetables is around 80 degrees F, but on a sunny day, the temperature inside the cold frame can reach 100 degrees F, so be sure to vent it appropriately. Conversely, keep the cold frame closed when the temperature dips below 65 degrees F: Cold winds will stress or kill these tender vegetables. Your heat-loving seedlings will be able to survive on their own two weeks *after* your spring frost date. This is a good time to disassemble your cold frame and store it until the fall.

Organizing the Fall and Winter Garden

One of the hardest things for me to figure out was how to save room for a fall/winter garden. Every inch of my small garden is continuously planted, and by late summer I have no desire to pull out perfectly productive plants to make way for the fall garden. After many years of not growing a fall/winter garden, I finally wizened up and figured out a plan.

What has been especially effective for me is to organize my garden by growing cold-loving and heat-loving crops in different beds. Grouping crops together in this way allows me to manage their needs more effectively, which means a better fall and winter garden for me. For example, a cold spell in late fall will be problematic for heat-loving summer vegetables like tomatoes and peppers. I can extend the season by covering them with row covers for a light frost, or make a last harvest before a heavy frost.

On the other hand, my cold-loving fall vegetables will thrive unadorned in a light frost. In fact, most cold-loving vegetables will taste better after the first light frost. This saves me time, because in the event of a frost, I don't have to cover every row of vegetables—only the heat-loving, cold-sensitive ones that are grouped together. I also save money because I don't have to own frost protection implements to cover the entire garden, only select areas.

Vinegar jug cloches

Plant cold-loving spring crops together, and continue to sow them regularly in the same bed throughout the entire season into fall, as I've outlined above in succession planting. This will ensure a harvest of vegetables straight through fall, and perhaps even into winter with the right protection. Sometimes the garden bed for these cold-loving crops will look empty in the height of summer when germination is low. Remember, daily waterings may be needed for successful germination, and low germination rates in hot weather are common. If several attempts of summer sowings aren't successful, just add some compost soil, mulch the bed well, and wait until 12 weeks before your first fall frost

date to begin sowing cold-loving crops for a fall/winter harvest. You can also try covering your newly seeded bed with summer weight row cover to help retain moisture until the seeds have germinated.

If you would like to try growing food year-round, row covers and cold frames are two ways to extend the harvest season into fall and winter. Row covers will provide two to four degrees F of frost protection, while cold frames generally offer seven to 10 degrees F of protection. Used together, row covers and cold frames can create an environment with a 9- to 14-degree F temperature difference for crops on the inside.

> DOUBLE COVERAGE MOVES THE COVERED AREA ABOUT THREE USDA ZONES TO THE SOUTH.
>
> *ELIOT COLEMAN, THE WINTER HARVEST HANDBOOK*

For example, a well-established broccoli plant is among the vegetables listed as being hardy down to 25 degrees F. This means that broccoli protected by a cold frame and row cover may be able to survive 16-degree F temperatures, and possibly even 11-degree F temperatures.

Many season extension micro-farmers take even more measures to ensure a winter harvest by surrounding their cold frames with straw bales for insulation, and/or covering the outside of the cold frame with old blankets. Each layer will increase frost protection.

To have a successful fall/winter harvest season, the importance of planting the crops at the appropriate time cannot be emphasized enough. Many fall and winter crops are sown as early as July. Check the planting schedule for the appropriate last planting dates for your area (download the **Seed Starting & Planting Worksheet** at www.TenthAcreFarm.com/tsmf-companion.)

General Tips for Cold Frame Success

Because you're altering the garden environment by using a cold frame, it's important to pay attention to your new vegetable starts to ensure they grow healthily. Here are some tips to keep in mind.

Organizing the Fall/Winter Garden - 50° F

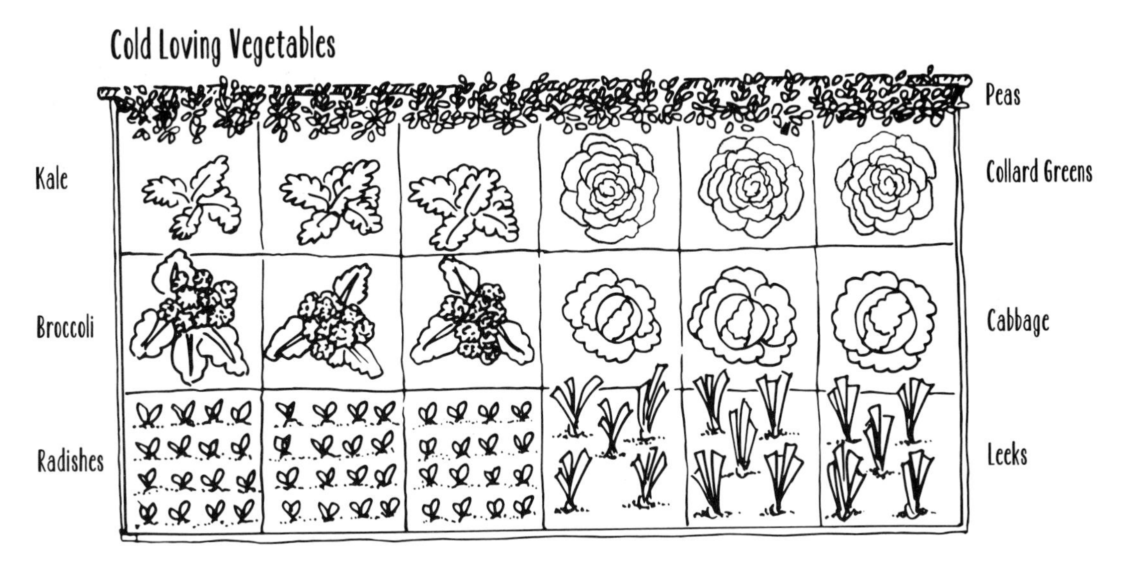

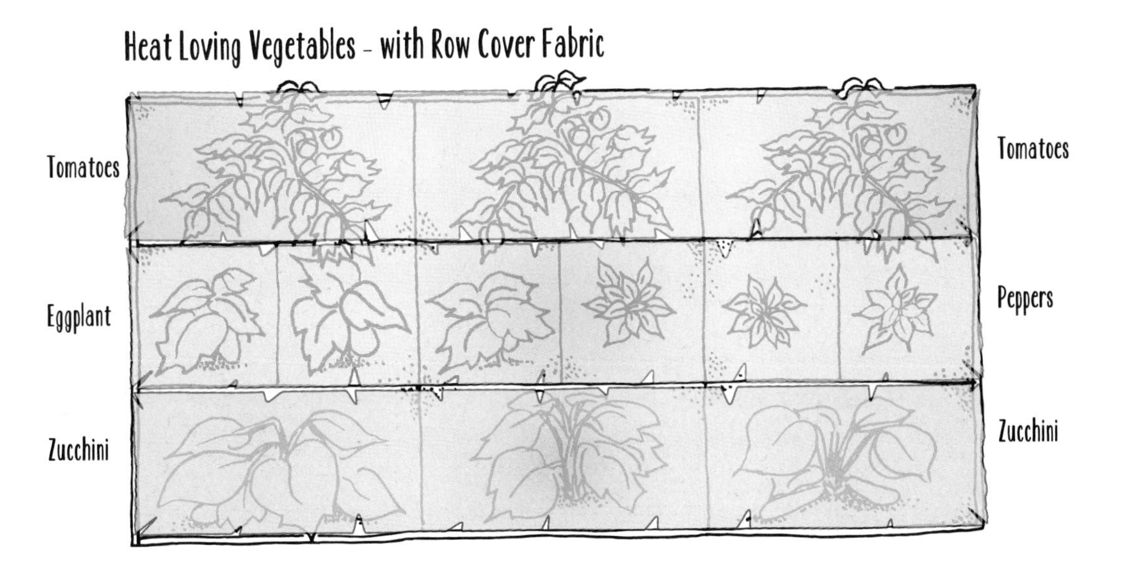

Organizing the Spring/Summer Garden - 90° F

Cold Loving Vegetables - with shade cloth

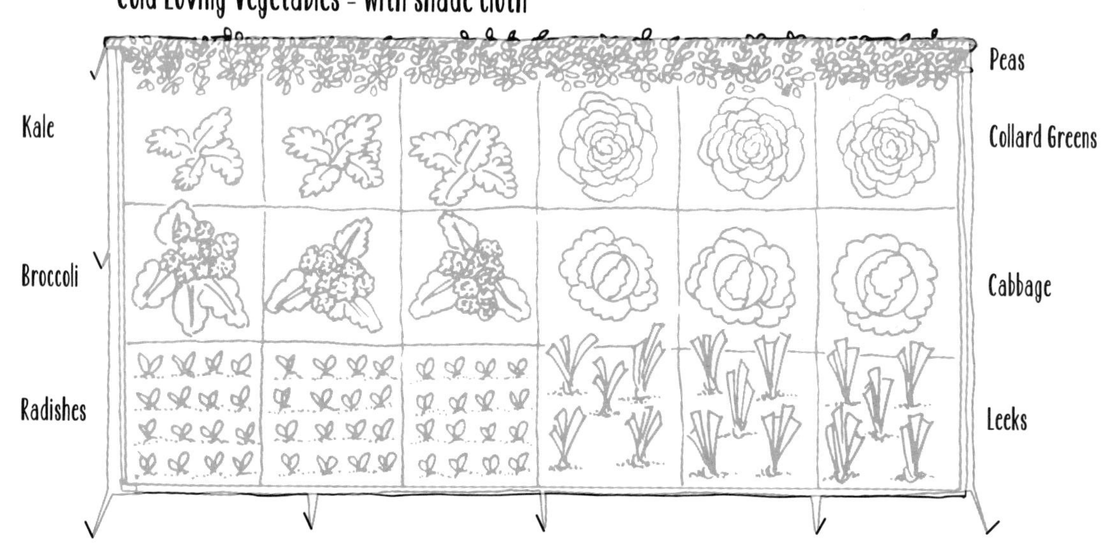

Heat Loving Vegetables

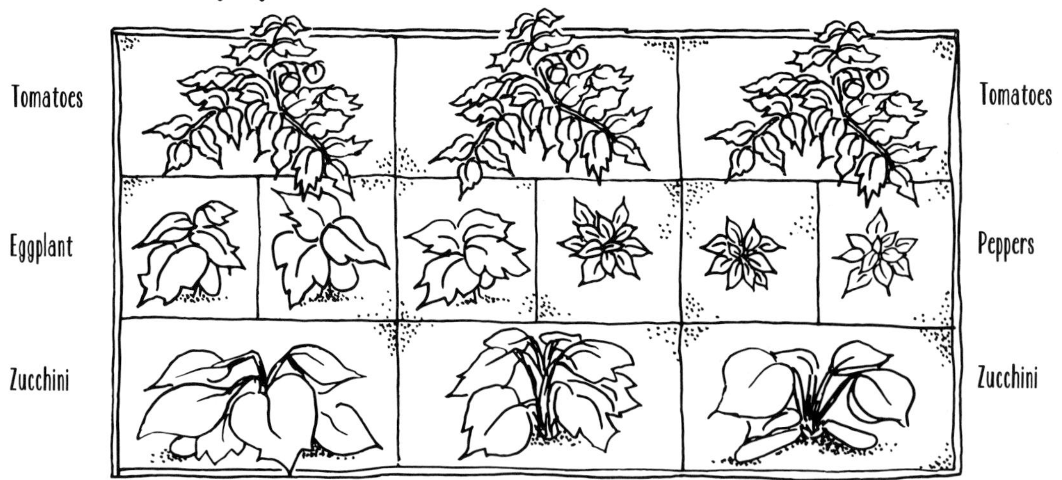

Water

Soil inside the cold frame should always remain damp, so check it frequently and don't allow it to dry out.

Ventilate

Ventilation is required to remove excess moisture and prevent disease, as well as regulate temperature. If you're unable to monitor the temperature daily, consider installing an automatic vent opener, which will automatically open the cold fame when temperatures rise to between 55 and 75 degrees F.

Bundle Up

If the outside temperature is expected to fall below 33 degrees F, you may need to add blankets on top of and around the cold frame to protect the seedlings.

Use Additional Season Extenders

Winter weight row covers and cloches on the inside can warm the soil an additional two to four degrees F. When used in conjunction with cold frames, vegetables will be well protected down to 23 degrees F. Each additional layer will increase frost protection.

Whether you're growing in spaces that are large or small, sunny or shady, or susceptible to pests or cold, there are always ways to make a micro-farm a productive and efficient system. Your best resource is your ability to adapt.

5 FRUIT ON THE MICRO-FARM

YOU'VE GOT TO GO OUT ON A LIMB SOMETIMES BECAUSE THAT'S WHERE THE FRUIT IS.

WILL ROGERS

Growing fruit at home is one of the most economical ways to get a return on your gardening investment. That's because most fruit crops are easy to grow, come back every year, and produce a high value harvest that saves you money. Plus, fruit is delicious! It can be a healthy snack on its own or transform into a decadent dessert.

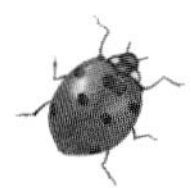

Growing Fruit Trees

A fruit tree can be the single-most source of fresh produce for the least amount of space and effort. As a bonus, trees improve biodiversity, and the roots will help to reduce erosion.

Fruit trees are best planted in the early spring or fall. Summer heat and dry weather can be taxing to a newly planted tree. In a hot climate, summer-planted trees need more water to get established, and rarely get enough to establish a strong root system. The cooler weather of spring or fall offers a reprieve from high temperatures, requiring far less water.

Are Dwarf Fruit Trees Right for You?

Dwarf fruit trees take up less space, are easier to manage and harvest from, and will produce fruit at an earlier age than regular-sized trees. Because dwarf varieties take up less space, more of them will fit into an allotted space, giving you a higher overall yield. They seem like a win-win, right? If you have good soil—or plan to amend it at planting time—and are available to water new trees regularly during the growing season, then dwarf fruit trees can be a good solution.

However, dwarf fruit trees are bred to have smaller root systems, which means that in nutrient poor and dry soils, their roots aren't extensive enough to reach for needed nutrients and moisture. If you plan on planting your fruit trees in a dry climate or a remote location where the soil won't be amended and the trees won't be watered throughout the growing season, standard-sized fruit trees (which can be pruned to a smaller size) will likely be a better choice.

Hardiness

Be sure to choose varieties that are appropriate for your USDA hardiness zone (see www.TenthAcreFarm.com/tsmf-companion for a link to a hardiness zone map).

Pollination

You may have seen a fruit tree with beautiful flowers that didn't produce any fruit. That's because many fruit trees require cross-pollination of its flowers to produce fruit. Cross-pollination is the

transfer of pollen from one variety of the same fruit tree to another variety. It means that some fruit trees—such as apples—require two different varieties to be planted within 500 feet of one another to have good fruit set. When purchasing trees, be sure to note when at least two varieties will be needed, and which varieties cross-pollinate one another.

Selecting a Fruit Tree Planting Site

Choose a location where your new fruit trees will receive eight hours of sun each day to minimize fungal infestations and to encourage full fruit production.

Protection from Winter Cold and Spring Frosts

Avoid planting fruit trees in low spots where cold air settles. A northern slope or the northern side of a building is a good location for fruit trees because the shadow from the building will protect the tree from harsh winds and protect young buds from a killing frost. In this shade, bud development and fruiting will be delayed, but it will provide more assurance of a successful future harvest. If your growing season is short, you may consider planting your trees on the south side of the building, but be aware that the tree will be more susceptible to frost and sun damage.

Good Drainage

Most fruit trees don't like wet feet, so make sure your planting site drains well. If water pools in the planting area, it should drain completely by 24 hours after the rain event.

Space

Each tree has an estimated width at maturity called crown diameter, also called the drip line. The space you choose should accommodate your tree's mature size, and trees should be given adequate spacing for proper air flow and easier harvesting.

TIP: PLANT A FRUIT TREE GUILD

A guild is a permaculture technique in which fruit trees are under-planted with herbs and ground covers that provide fertilizer and mulch, attract beneficial insects and pollinators, and deter pests. To read more about fruit tree guilds, see chapter 11.

This apple tree was planted with comfrey leaves added to the planting hole. Fresh herbs of all kinds (without their flower heads attached) will act as a slow-release fertilizer.

Plant a Fruit Tree in Three Steps

After you have chosen a variety of fruit tree that is appropriate for your hardiness zone and selected a planting site that is optimal, you're ready to plant.

1. ***Dig a hole.*** Dig a hole twice as wide and deep as the roots of the young tree.
2. ***Place the tree and top with compost.*** For this step, it's helpful to have two people. Hold the tree in place at soil level while fanning the roots out like an umbrella. Make sure the trunk of the tree is straight. Another person backfills the hole with soil, breaking up any clumps and building a solid foundation underneath the crown. Cover all roots, and ensure the graft union is above soil level. Lightly tamp down the soil with your foot to remove any air pockets. Now spread a layer of compost on top of the soil. As it rains, the compost will fertilize the new tree.

3. ***Prune and water.*** Cut the tree back by one-third to two-thirds of its existing size (when planted in spring or fall). It might seem harsh, but this action will actually encourage more prolific growth of the root system, which is essential for a strong and resilient tree. Also prune any dead branches and water the tree in well. Continue to water it well at least once per week throughout the growing season.

Plant a Windbreak

Consider using a hedgerow around your orchard area to protect your fruit trees from wind, attract beneficial insects, and lure birds away from your fruit. Birds especially love elderberries, mulberries, ornamental cherries, and crabapples, and planting them nearby will keep them from wanting your fruit harvests. See chapter 11 to read more about hedgerows.

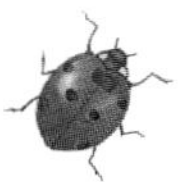

Growing Berries for Beginners

Berries are a high-value, nutrient-dense crop, and berry bushes will grow with minimum care. Bonus: if you don't have a chance to harvest the berries, they'll feed the local wildlife!

Planting any edible perennial requires patience, since they usually take from one to three years to begin producing a worthy yield. Luckily, there are a few quick-yielding crops that will help tide you over while you wait.

Red raspberries

Quick-Yielding Berries

Ground cherries (*Physalias* spp.) are an annual in the nightshade family that will produce heavy harvests like other nightshade vegetables such as tomatoes. Because they are frost-sensitive, they will need to be replanted each year. Ground cherries are also called husk cherries, strawberry tomatoes, or husk tomatoes. They are golden, tart-sweet berries

with a strawberry-pineapple like flavor, reminiscent of cherry tomatoes except that they grow inside a paper-like husk. Ground cherries can be eaten fresh and are popularly used to make jams, preserves, baked goods, or salsa.

Raspberries are the third most popular fruit in the United States for fresh eating, behind strawberries and blueberries. Some raspberries produce two crops, one in spring and another in fall, called everbearing or fall-bearing. Everbearing raspberries that are planted in the spring will usually produce a harvest in the fall of the first year. No waiting!

Ease into It

Some of us need to stick to the basics—those berries that are easy to care for—while we're getting our feet wet with farming. Blueberries and blackberries fit the bill, and are sweet, juicy, and low-maintenance.

Blueberry bushes are beautiful shrubs that are slow-growing and take about three years to begin bearing fruit. If you're busy setting up a new homestead, this is one crop that you can plant right away and then get on with other tasks while they get established.

Blackberries, like raspberries, are considered brambles, which are prickly shrubs made up of stems called canes. They are easy to grow once you get the hang of the pruning routine, and the large, juicy berries will yield their first harvest in the second summer.

All of these berries are considered soft fruits, which means that they will last only a few days in the refrigerator. Micro-farmers will want to have a plan in place to process the berries soon after harvesting, either by fresh eating, baking, canning, dehydrating, or freezing.

Here's what you need to know to get your berries off to a good start.

Ground Cherries

Start ground cherries as you would tomatoes, indoors six to eight weeks before your average last frost date. Transplant out after all danger of frost has passed. Plant them deeply and in full sun - like tomatoes - about three feet apart. The bushy plants will grow three feet tall and can be trellised using tomato supports, or they can be allowed to sprawl on the ground. Harvest them when the papery husk begins to turn light brown and folds back to reveal the ripened fruits.

Ground cherries are not frost hardy. Try 'Aunt Molly's,' a heritage variety from Poland known to be prolific and super sweet. Practice good crop rotation, as ground cherries are susceptible to many of the same diseases as other nightshades.

Raspberries, blueberries and blackberries are all perennial fruits. With a single planting comes the single chance to set up their site properly. Clear the planting area of weeds and work in compost and other soil amendments before planting. Spring and fall are the best times to plant perennials.

Everbearing Raspberries and Blackberries

Choose a site in full sun with good drainage. Raspberries will enjoy a north-facing slope or the north side of a building in order to protect them from late spring frosts. Raspberries are self-fruitful and produce fruit without cross-pollination. Space raspberry plants two feet apart and blackberries 2 ½ feet apart in rows. Keep them well-watered in their first year. Use a trellis system to keep canes and berries off the ground, and to make the berry canes more manageable. After the canes produce berries in the fall of the first year, they will die. Prune these canes back either by hand or by mowing.

Blueberries

For the best harvest, choose a site that is in full sun. Blueberries will enjoy a north-facing slope or the north side of a building in order to prevent damage from late spring frosts. It is essential that the site be well-drained. Blueberries have specific soil needs: the pH should fall between 4.5 and 5.0, and soil must be amended before planting if it doesn't meet this requirement. Blueberries grow well on slopes or in raised beds where drainage is ideal. They can be planted in either spring or fall, spaced about four feet apart.

Keep them watered and mulched—wood chips or sawdust is best. No pruning is required in the first year. Cross-pollination is not required, but you'll have better success with more than one variety.

Strawberries

Planting and Growing Strawberries

Strawberries are the darlings of the berry world, and homegrown strawberries are a popular choice for micro-farmers, not only because of the quality and freshness, but also because of the health benefits and cost savings. They are quick to yield compared to many other fruits: the first harvest will come in the late spring of their second year.

Growing strawberries is a wise choice for the health-conscious consumer. After analyzing pesticide residue on 48 popular produce items, the Environmental Working Group ranked

strawberries fourth-highest in the amount of pesticides present, just behind apples, peaches, and nectarines. Purchasing chemical-free strawberries from a local farmer can run upwards of $5 a quart, while growing your own would cost less than 50 cents per quart over three years. Berries are an economical choice for the frugal micro-farmer.

So let's plant some strawberries, shall we?

There are two main types of strawberries: June-bearing and everbearing. June-bearing plants are the most popular to plant, producing large yields in early summer, hence "June." They also produce a lot of runners, called daughter plants. Plant June-bearing strawberries 12 inches apart.

Everbearing strawberries, on the other hand, are smaller and produce slightly less yields at one time, but they produce two crops, one in late spring and the other in early fall. Everbearing strawberries also produce fewer runners, which means that they are the better choice for the edible landscape. I have them growing in my front yard! Plant everbearing strawberries eight inches apart.

Year 1:

Plant in spring. Remove runners and flowers throughout the year.

Year 2 and 3:

Harvest berries late spring / early summer.

Year 4:

Harvest berries. After harvest, remove plants and compost them. Add aged manure and seed with clover.

Year 5:

In early March, cut the clover back to the soil line. Let sit on top of the soil for 2 days, then incorporate it into the soil with a digging fork, breaking up the roots. After 3 weeks, plant new strawberry plants. Begin again by removing runners and flowers.

Growing Strawberries

Early spring is the best time to plant strawberry plants so they have enough time to develop roots before winter. Strawberries can be planted from four weeks before to four weeks after your spring frost date. Choose a date when the ground is not too wet.

Strawberries do well in a raised bed or on a slope with good drainage. When planting, the soil should be loose enough that the roots can set vertically. Do not bend roots sideways. Half of the crown should be planted above the soil, while the other half is planted below. Do not allow roots to be exposed above the soil, and never cover the entire crown.

In the first year, remove runners and flowers to encourage the plants to put energy into developing strong root systems. This will aid in a larger harvest in the spring of the second year. Keep plants well-watered in the first year.

Strawberry plants produce their highest volume in years two through four. Strawberries are perennial plants, but their yield and quality decline after three harvest seasons, and will need to be replanted. After the third harvest in the fourth year, remove plants from the bed and compost them. Add an amendment such as manure, and seed with a non-grass cover crop. This will replenish soil fertility and discourage pests. In the spring of the following (fifth) year—about 23 days before planting new strawberry plants—cut the cover crop back at the soil line. Let the cover crop plant matter sit on top of the soil for two days, then incorporate it into the soil with a digging fork, breaking up the roots. After three weeks, the bed can be planted, and you'll start again by removing runners and flowers. To avoid having a season without a strawberry harvest, start a second bed in year two.

With my first strawberry bed, I didn't dig up the plants after the third harvest season. I just couldn't imagine digging up plants that were producing strawberries so vigorously. I had grown them almost completely pest-free for the first three years. However, I discovered that the recommendation was in fact a wise one. In the fourth year, my strawberry plants produced almost no berries, and ants and slugs damaged the berries that they did produce.

Red currants almost ready for harvest

Growing Currant Bushes

The first step we took in transforming our front yard into an edible landscape was to replace the traditional hedge lining our front porch with currant bushes. This area on the north side of the house is almost completely shaded.

Currants are thornless shrubs that produce red, black, or white berries. We planted both red and black varieties in our front hedge. Although our hedge is beautiful, if I were planting it again, I would have chosen just one variety to achieve more consistency.

Currants are understory bushes that naturally grow in dappled shade. In shade, they'll produce lusher, darker foliage, which is good news for an edible landscape. However, currants can often produce more fruit in the sun; they'll just require more water to thrive there.

TIP: LANDSCAPE LIKE A PRO

Choose only one variety of a species to serve as a foundation hedge. For more landscape design ideas, see chapter 10.

Ways to Use Currant Bushes in the Landscape

1. As the main ornamental hedge, such as ours bordering our front porch.
2. In a partially shady, unused spot, particularly at the drip line of oak, walnut, or apple trees.
3. In a wildlife hedge. Deer and birds love currants.
4. In a poultry foraging area. Chickens like currants, too.
5. At the edges of open woods and in dappled shade in the woods.
6. In a pollination garden. The tiny flowers provide nectar for both hummingbirds and a menagerie of other insects.
7. In your medicinal garden. Black currant leaves have been used traditionally to support joint health and promote healing from common respiratory illnesses.
8. In a jelly garden. Transform these tart (red) and musky (black) berries into sweet and mellow treats. Read more about jelly gardens later in this chapter.

In my opinion, the black currant bushes are a lot prettier than the other varieties. The bushes are more full and upright with straighter branches, and grow to about four to five feet tall and wide. The leaves are bigger than those of red currants and the branches provide more interest in the winter because the red buds are bigger. The black currant bushes are also twice as productive as the red currant bushes. The red currant bushes are smaller (three to five feet tall and wide) and

a little more scraggly looking. But that's kind of a neat look, too, with the gnarled branches. Both need to be pruned in the late winter for good berry production, and they can be pruned into a tidy hedge shape if that's your thing.

Red currants are ready to harvest in late May to early June, while the black currants come into harvest about two or three weeks later. They're both harvestable for about a month. When the entire bunch of berries is ripe (like a bunch of grapes), harvest the whole bunch. Once harvested, pick each berry off the stem before eating or processing.

I freeze both red and black currants and use them in fruit smoothies mixed with other berries. I also use the frozen berries to make mixed-fruit jelly. Black currants dehydrate well. The dried berries can be used in baked goods, desserts, granola, and tea. I use the red currants to make a berry infused vinegar, which is delicious on vegetable salads, fruit salad, or as a marinade for fish, pork, lamb, duck, or chicken. I make a berry infused vodka with the black currants, which becomes a French liqueur *crème de cassis* when sugar is added. It makes a delightful cocktail called *kir* when mixed with white wine or champagne.

Bee on currant flower

Growing Black Raspberries

Black raspberries also do well in part shade. As I've mentioned before, we replaced the hedge lining the front of our house with a row of black raspberries. The berries are so delicious that the harvest rarely makes it into my kitchen! They have a richness of flavor that I find more

Red currant vinegar

enticing than red raspberries. And although there is a lot of information to be found on growing red raspberries, black raspberries are less talked about. Interestingly, red and black raspberries have different growing habits and cannot be managed in the same way. Black raspberries, like their red cousins, are highly perishable, which leads to a high price at the grocery store. This makes them an economical choice for the landscape.

Black raspberries are a beautiful landscape plant. Bright red canes blaze confidently through gray winters; pinks, reds, and purples of the ripening berries are beacons of cheer in early summer. Unlike red raspberries, black raspberries have a clumping habit. New canes emerge each year from the original root crown, which means that black raspberries "stay put"—a good trait to have in a small-scale landscape.

Black Raspberries Lend Themselves to the Small-Scale Landscape

- Their height never exceeds 2½ to 4 feet (if they're pruned properly).
- They're thorny and could provide some security near windows.
- They're beautiful in both winter and summer.
- They "stay put."
- They're shade tolerant and grow well in areas that are overshadowed by a house or other building.

The Downsides to Growing Black Raspberries

1. Birds love them. In fact, out of all of the fruit in our yard, black raspberries are by far their favorite. Bird netting and shiny things like old CDs will deter them somewhat. To better deter the birds, prune the canes to a shorter height (around three feet), where the birds will not risk contact with ground predators such as cats.
2. Deer love them, too. Fencing black raspberries will be your best defense, as deer will eat the entire plant, not just the berries. For suggestions on growing fruit in deer country, see chapters 10 and 11.

Plant black raspberries on the north side of a building to keep them protected from frost and wind damage. Choose a location in full sun or one that is partially shaded. Plant the canes 2½ feet away from each other in a row. Be sure you can access both sides of the row for harvesting, training, and pruning. Plant black raspberries at least 300 feet from wild raspberries or

blackberries, which can spread disease to your black raspberries.

At the same time as planting, build a supportive trellis-type system. Be aware that support structures are different for black raspberries and red raspberries.

The area where you've planted an original cane is called a *hill*. Set a 2x4 or 4x4 post behind each "hill," 1½ feet deep. The set post should be about four to 4½ feet high. Setting the posts at the time of planting ensures the least amount of stress on the plant's root system. Secure a heavy gauge wire or strong twine to run from post to post near the top. Creating a notch in the wood for the wire or twine to sit in allows it to stay secured.

As the canes grow, train them to drape over and run along the wire or twine in between the hills.

Black raspberries ripening

Pruning Black Raspberries

Regular pruning will result in larger berries and larger harvests. It will also keep the berry patch manageable and more space-efficient. If you have an old black raspberry patch, skip to the directions for *Renovating an Old Black Raspberry Patch*.

Note: Pruning requires leather gloves and covered skin. Long sleeves, pants, and closed-toe shoes make this task more enjoyable.

In the first year, do nothing except admire your plants' energetic will to live! After fruiting—and before winter—canes are headed (pinched, tipped, or cut off) at a desired height, around 28 to 30 inches. (A cane emerges directly out of the ground.) Loosely tie all of the canes in each hill to their coordinating post. This will help to keep walkways clear of thorns.

After making it through the winter, a spring pruning will ensure a fantastic summer harvest. When the plants are beginning to create buds—and before the plant leafs out—remove dead canes. Canes that produced berries in the previous year will be dead, so cut them back to the ground. Other canes will have been damaged by the cold and will be brown and dead. Cut all dead canes as close to the ground as possible.

Regular pruning of a black raspberry patch will keep it manageable.

Thin the canes to the four to six strongest canes per hill. Cut the rest back to the ground. If your plants are young and haven't produced this volume of canes yet, then you can skip this step.

In the previous fall, each cane was cut back to a desired height, which inspired a lot of side branches—or laterals—to grow. These laterals are where the berries develop. In the spring, for each lateral, count eight to 10 buds away from the cane and then cut the rest of the long branch off.

Laterals are quite the weaving, tangled mess in the spring. Between removing dead canes, thinning, and heading back the laterals, you will take quite a bit of biomass away. Black raspberry plants will look dramatically different after pruning, but have faith that the yields will be better. You may again wish to loosely tie the canes in each hill to their coordinating post to keep everything tidy and out of walkways.

Renovating an Old Black Raspberry Patch

The process for renovating an old black raspberry patch can be started in either spring or fall. This will not be fun! There are thorns galore in an unmanaged patch.

Any time after the fruiting season is over—and before winter—cut all of the canes back to 28 to 30 inches tall. Measure to be sure that the center of each hill is around 2.5 feet apart in a straight row. If unmanaged, many of the canes will have bent over and rooted themselves outside of the hill. If the row is crowded, dig up the extra canes and find a new home for them.

Spring is the best time to construct a support system if your berry patch doesn't have one. While the posts are typically set just a few inches back from the hill, the posts for a renovated berry patch may be set about six inches to one foot behind each hill, in order to avoid piercing the bulk of the root system. You could also consider an "X" pole solution (two six-foot plant stakes installed in an X shape). They will not pierce as much of the root system. Run wire or twine between each hill.

Black raspberry harvest

In the spring, for each lateral, count eight to 10 buds away from the cane and then cut the rest of the long branch off.

If you just started your renovation in the spring then follow this modified plan: Remove dead canes by cutting them off at the base. Next, thin the canes to the four to six strongest canes per hill. Cut the rest back to the ground. Now, drape the long branches over the heavy gauge wire in the support system. You should get an improved harvest this year, but be sure to follow a regular pruning schedule for an even better harvest next year.

Harvesting Black Raspberries

Black raspberries take about three years to get a full harvest. Learn about the ripening season for the specific variety you plant. Different varieties will ripen in early-, mid-, or late-season. Ours ripen in June, so June is not the month for us to go on vacation if we want to harvest black raspberries! When the berries are ready to harvest, they'll turn from bright red to deep purple. Once they begin to ripen, harvest every day for the peak of freshness and to beat your wildlife friends (who will inevitably get some anyway).

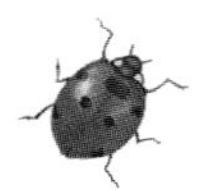

Choosing Fruit for Your Circumstances

We all wish that we had the perfect, full-sun garden with amazingly fluffy, rich, well-draining soil. Alas, because nature if perfectly imperfect, so are our growing spaces. I think the universe simply wants to spark our creativity with garden design challenges.

Fruit for Wet and Erosion-Prone Areas

Since humans don't typically enjoy wet feet, we have adapted to cultivating and eating fruit that naturally grows in well-drained areas. But don't let that fool you. A decent number of edible berries can be found in soggy areas. For example, elderberry, gooseberry, and highbush cranberry are fruit-producing shrubs that don't mind temporary wet feet. They could even be planted in a rain garden that drains within 24 hours. For more wet-area crop ideas, see chapter 11 for "Edible Crops in Rain Barrels".

Fruit for Shady Areas

Shade can disappoint even the most enthusiastic gardener, but it doesn't have to mean growing fruit is a no-go. In fact, most fruit-producing species evolved in part-shade environments. In addition to currants and black raspberries, other fruits that will grow exceptionally well in partial or dappled shade are elderberry, gooseberry, serviceberry, and pawpaw.

Grow a Jelly Garden

Many of us purchase properties that include at least one tree or shrub that produces seemingly pointless fruit: a crabapple or Bradford pear tree, for example. Privacy hedges often include many shrub species that produce an edible—but rather unappealing—berry. If you have a hankering for a productive landscape, you might wish to tear out that ornamental crabapple and replace it with a real apple tree, or replace the hedge with a tastier edible species. This sentiment is understandable, and in small yards where the one tree or hedge is the only opportunity to grow edible fruit, I might do just that.

However, you might be surprised to know that crabapples, Bradford pears, and many shrub-produced berries are actually edible. With some creativity, you can learn to appreciate the abundance of fruits that produce without any help on your part.

Caution: If you decide to harvest from existing plants, be sure you have accurately identified the plant to avoid poisonous berries.

Pollinators love crabapple blooms when they wake from winter slumber.

Red currant harvest

Tart-sour cherry harvest

A black currant harvest

While many edible ornamental and wild fruits are overly tart and seedy when eaten fresh, cooking them into jelly transforms them from tart and seedy to sweet, mellow, and smooth. Enter the jelly garden. Kids could really get into a project like this, too. Mixing and matching fruits and their flavors to create the best jelly would be a lifelong journey of getting to know a place: a labor of love and a declaration of your *terroir*.

Your very special and unique jellies could be a great source of extra income or gifts: Who wouldn't be curious about the mini Bradford pears or crabapples from those ubiquitous neighborhood trees turned into a delicious jelly, for example? Both are usually harvested in the fall, just in time for holiday jelly making.

Autumn olive is another example of a tree that is widely dispersive (read: a nuisance) with free berries for the picking. Harvest the berries, make a delicious jelly, and keep fewer seeds from spreading!

There is an almost infinite number of edible, berry-producing plants, but if you're planting a jelly garden intentionally, here are some ideas to get you started. All of these berries pack a powerful, nutrient-dense punch. Delicious jelly is only half of it: These shrubs and trees will also create biodiversity and benefit your local ecosystem.

Currants–Ribes spp. *(hardiness zones 3-8)*

Currants are harvested in bunches, like grapes. Black and red currants are the most common varieties cultivated for their edible berries. Musky and tart respectively, these seedy berries become a delight when cooked.

Currant flowers are a preferred source of nectar by both hummingbirds and beneficial insects. Songbirds and chickens both enjoy the fruit, too. Currant bushes can be planted in a deer-deflecting hedge to keep deer out of certain parts of the yard, and they can also be useful in a windbreak hedgerow. They are even juglone tolerant and can be planted with walnut trees. See the "Growing Currant Bushes" section earlier in this chapter.

Elderberries–Sambucus spp. *(hardiness zones 3-10)*

Elderberries are a summer berry that is ripe for picking when it is dark purple and soft. Elderberry bushes can grow in many kinds of environments, but they will do well in wet, low-lying areas and on the banks of waterways. They are juglone tolerant and can be planted with walnut trees. Elderberries and their flowers have useful medicinal properties.

They are beloved by both songbirds and chickens—when you see the birds going for the berries, they're ripe! The shrubs are a popular wildlife shelter and nesting site. Beneficial insects and hummingbirds are attracted to the nectar of the edible flower.

Raw elderberries can upset the stomach in some people, so cook before eating.

Mulberries–Morus spp. *(hardiness zones 5-9)*

Mulberries can be white, black, or red. Make sure the fruit is soft and ripe before picking, as unripe fruit can upset the stomach. To ensure picking only ripe fruit, place a sheet on the ground underneath the tree and shake the tree to drop the ripe fruit.

Mulberries are beloved by birds of all kinds. Songbirds will often prefer these summer berries to tree fruits (saving your more valuable harvests), and chickens and ducks also enjoy the berries. The trees, which are fast growing in disturbed areas, will quickly become shelter and nesting sites for birds.

Mulberry trees are also known to be tolerant to juglone and can be planted with walnut trees.

Rose Hips–Rosa rugosa *(hardiness zones 2-7)*

Rose hips have 60 times more vitamin C than citrus fruits. Harvest the red berries after the first frost.

The old timey rosa rugosa is a heavenly-scented rose. Unlike its cousin—the modern, hybridized rose that lacks pollen and needs a ton of maintenance to stay healthy—rugosa rose is a low-maintenance shrub and important nectar source for beneficial insects and hummingbirds. It provides shelter and nesting sites for birds and other wildlife and winter fruit for the birds. It is even known to be juglone tolerant and can be planted underneath walnut trees.

Tart Cherries–Prunus cerasus *(hardiness zones 4-8)*

Tart cherries are excellent for fresh eating when ripe in my opinion, but many people find them extremely tart. They mellow when pitted and cooked.

Cherry trees of all kinds are a popular insectary during their spring bloom time, and the berries are a favorite wildlife food. Cherry trees make a beautiful privacy screen. Each year, we harvest about nine pounds of cherries per tree.

Plant a Hedgerow Jelly Garden

If you were thinking about planting a jelly garden on purpose, it would look fantastic grown as a hedgerow, which is a narrow strip of mixed plantings. A small hedgerow would be 6′-x-12′ or 10′-x-20′, for example. See chapter 11 for more ideas on planting your jelly garden as part of a hedgerow.

Making Jelly

It is entirely possible to mix and match all of these berries (or any other edible fruits you've foraged) to make a mixed-fruit jelly. Simply follow the guidelines on the package of your pectin. If you prefer to experiment with each berry separately, the pectin box will have you covered with that, too, but any preservation recipe book will do.

Mixing fruits, rather than using a single berry, will give the jelly more depth of flavor. One year, I made a mixed berry jelly with red currants, black currants, and black raspberries. It was divine! Some people prefer jam to jelly for the same reason—depth of flavor. Jam includes the whole fruit, which creates a richness of flavor. However, in this case, I prefer jelly because I don't like the seeds.

Because most of these berries can perish quickly, it will be important to either make jelly right away after harvesting or preserve them for later. I like to freeze berries on the spot, and thaw them later when I have time to make jellies and jams. The only berries that will not freeze well are the rose hips, so if jelly is your preferred preservation method, you'll want to make rose hip jelly right away. Or you can dry them for use in teas. Black currants and mulberries can be dehydrated as an alternative to freezing.

Aside from peanut butter and jelly, some of my favorite ways to use jelly are: in thumbprint cookies, mixed in plain yogurt with granola, or on pancakes.

Using these "pointless" berries for jelly is just the tip of the iceberg. Use the frozen berries (mixed!) in smoothies. Try the dried berries to make chocolate-berry-nut bark, in granola, in baked goods, and in teas. Fresh or frozen berries can be turned into fruit syrup, pies, tarts, and other pastries. Add the unique taste and nutrient content of "jelly" fruits to your diet and grow plants that benefit wildlife, too!

Whether you're growing a whole orchard of fruit trees, a berry bush hedge, or foraging wild fruits, there are plenty of opportunities to add fresh fruit to your diet and kitchen repertoire.

HERBS ON THE MICRO-FARM

MUCH VIRTUE IN HERBS, LITTLE IN MEN.

BENJAMIN FRANKLIN,
POOR RICHARD'S ALMANAC

Herbs are easy to grow, nutrient dense, prolific, and low-maintenance, making them an excellent crop for the micro-farm. Even in a small garden, herbs can provide many of a household's culinary and medicinal needs. In addition to a nice harvest, some herbs make beautiful cut flowers or yield heavenly scents. Still others help fertilize the soil, provide mulch, attract beneficial insects, and deter pests. I like to choose herbs that have a many uses. Here are my favorite culinary and medicinal herbs, and those that improve biodiversity in the garden.

Kitchen Garden Herbs

It doesn't take many herb plants to provide most—if not all—of a household's fresh herb needs, which is a great return on investment considering the expensive price of small amounts of fresh herbs in the grocery store, and what little maintenance the plants require. It's easy to dry herbs and spices to use throughout the year.

Basil *(Ocimum basilicum)*

'Genovese' basil leaves are often used in Mediterranean and Asian cooking. To start indoors, plant seeds ¼-inch deep in seed-starting medium four weeks before your frost date. Or transplant/sow seeds outside, 10 inches apart, four weeks after your frost date. Basil is an annual that will need to be planted every year. It prefers a sunny, well-drained location but will do fine in partial shade.

Chives *(Allium schoenoprasum)*

Chives refer to the leaves of this onion-related herb. To start indoors, plant seeds ¼-inch deep in seed-starting medium 10 weeks before your frost date. Transplant outside, eight inches apart, the week of your frost date. Chives are a perennial that will come back every year in the same place. They need fertile, well-drained soil in a sunny or partially shady location.

Cilantro/Coriander *(Coriandrum sativum)*

Cilantro refers to the leaves of the plant, while coriander refers to the seeds. Its culinary uses range from the Mediterranean, North Africa, and Asia, to Latin America. To start indoors, plant seeds ¼-inch deep in seed-starting medium four weeks before your frost date. Or transplant/sow seeds outside, six inches apart, four weeks after your frost date. Cilantro is an annual that will need planted every year. It needs well-drained soil in a sunny or partially shady location.

Dill *(Anethum graveolens)*

Dill leaves are often used in flavoring pickles, salad dressings, and fish dishes. Sow seeds ¼-inch deep outside, 16 inches apart, two weeks before your frost date. Dill is a biennial plant that will come back a second year, and will often self-seed. It needs rich soil in a sunny location.

Marjoram *(Origanum majorana)*

Marjoram leaves are a milder form of oregano, popular in western European cuisines to flavor meat dishes. To start indoors, plant seeds ¼-inch deep in seed-starting medium six weeks before your frost date. Transplant outside, 15 inches apart, two weeks after your frost date. Marjoram is a perennial that will come back every year in the same place. It prefers well-drained soil in a sunny or partially shady location.

Oregano *(Origanum vulgare)*

Oregano leaves are popular in Italian, Greek, and Mexican cuisine. To start indoors, plant seeds ¼-inch deep in seed-starting medium six weeks before your frost date. Transplant outside, 15 inches apart, two weeks after your frost date. Oregano is a perennial that will come back every year in the same place. It prefers well-drained soil in a sunny or partially shady location.

Parsley *(Petroselinum crispum)*

Parsley leaves are common in Mediterranean and Eastern European dishes. To start indoors, plant seeds ¼-inch deep in seed-starting medium eight weeks before your frost date. Transplant outside, 10 inches apart, two weeks after your frost date. Parsley is a biennial that will come back a second year. It needs fertile, well-drained soil in a sunny or partially shady location.

Rosemary *(Rosmarinus officinalis)*

Rosemary leaves are often used in Mediterranean cooking and to season meat. Transplant seedlings outside, eight inches apart, four weeks before your frost date if you live in hardiness zones 1 through 7. If you'd like to keep your rosemary plant year after year, consider planting it in a pot and bringing it inside for the winter, because it may be destroyed by frost in cold winters. If you live in hardiness zone 8 or higher, rosemary can be planted as a perennial that will come back year after year. Outside, space the plants 18 inches apart. It needs well-drained soil in a hot and sunny location. Allow the soil to dry out in between waterings.

Sage *(Salvia officinalis)*

Sage leaves are popularly used in sausages and stuffing. To start indoors, plant seeds ¼-inch deep in seed-starting medium six weeks before your frost date. Transplant outside, 12 inches apart, two weeks after your frost date. Sage is a perennial that will come back every year in the same place. It needs well-drained soil in a hot and sunny location. Allow the soil to dry out in between waterings.

Thyme *(Thymus vulgaris)*

Thyme leaves are used in Mediterranean cooking in dishes ranging from vegetables and meats to eggs. Transplant seedlings outside, 12 inches apart, two weeks after your frost date. Thyme is a perennial that will come back every year in the same place. It needs well-drained soil in a hot and sunny location. Allow the soil to dry out in between waterings.

Growing a Medicine Garden

There are oodles of herbs that are easy to transform into healing remedies in the home. The following are a few of my favorites that are easy to grow and contain valuable medicinal properties. Learn to use these herbs to make poultices, salves, tinctures, elixirs, vinegars, and more. I am not a doctor and the FDA has not evaluated the following claims about the traditional and medicinal benefits of herbs.

Comfrey *(Symphytum x uplandicum)*

Comfrey has been used to promote the healing of acne, bruises, and scrapes. Russian comfrey (*Symphytum x uplandicum*) is preferable to true comfrey (*Symphytum officinale*) because its seeds are sterile and, therefore, will not seed itself around the garden. To plant Russian comfrey, purchase live root cuttings and plant them two feet apart—crown buried—in the ground as early as four weeks before your spring frost date, and until four weeks after your fall frost date. Comfrey is a perennial that will come back year after year. It prefers rich, moist soil in a well-drained spot with full sun or partial shade.

Dandelion *(Taraxacum officinale)*

Dandelion has been shown to support healthy liver and kidney function, corrects blood pressure imbalance, and can encourage the healing of skin ailments like acne. It sounds odd to plant

Basil makes a nice border in the garden.

Chives also make a nice border in the garden or edible landscape.

Marjoram, parsley, and fennel line this garden path.

Sage

dandelion because it is so readily available as a weed. However, this weed most readily populates areas where heavy foot traffic exists and harvesting safely in those areas is questionable. Dandelion seeds are actually available for purchase! Sow seeds ¼-inch deep outside, four inches apart, two weeks before your frost date. Dandelion is an annual plant that will self-seed. It prefers well-drained soil with full sun or partial shade. If you produce more than you can use, take heart that dandelion leaves go for a high price to culinary chefs.

Echinacea *(Echinacea purpurea)*

Echinacea is used as an immune stimulant, and the tea is often gargled for a sore throat. Sow seeds in the fall in a well-prepared bed. Or transplant seedlings outside the week of your spring frost date, two feet apart. Echinacea is a hardy perennial and prefers full sun or partial shade in rich soil.

Fennel *(Foeniculum vulgare)*

Fennel can stimulate appetite, support healthy digestion, and be gargled as a tea for a sore throat. Sow seeds ¼-inch deep outside, 16 inches apart, two weeks before your frost date. Fennel is a perennial plant in zones 5 through 10 and an annual in northern climates. It will also self-seed. Fennel needs rich soil in a sunny location.

Garlic *(Allium sativum)*

Garlic aids immunity, supports healthy blood pressure, and is traditionally used in remedies to eliminate common intestinal parasites. Garlic is planted in the fall starting six weeks before—and until—the fall frost date. Plant garlic cloves five inches apart and two inches deep, with the pointed end facing up. Mulch the bed well. Harvest in early summer. It prefers rich, moist soil in a well-drained location with full sun or partial shade.

German Chamomile *(Matricaria recutita)*

Chamomile encourages stress relief, promotes healthy digestive function, and restores restful sleep. Scatter the tiny seeds outside in a prepared bed, six inches apart, the week of your frost date. Gently tamp down the seeds with your hand to make contact between the soil and the seeds. German chamomile is an annual plant but will often self-seed. It enjoys partial shade and dry soil.

Echinacea

Lavender is sensitive to harsh weather. Here it absorbs thermal heat from rain barrels in spring and summer.

Peppermint grows in a pot on the shady front porch.

Healing salves are easy to make with homegrown herbs.

Lavender *(Lavandula angustifolia)*

Lavender has traditionally been used to support mental wellness, encourages germ-free environments, promotes healthy digestion, and corrects muscle tension. Transplant seedlings outside, 15 inches apart, four weeks after your frost date. Lavender is a perennial that will come back every year in the same place. It requires well-drained soil in full sun. Allow the soil to dry out in between waterings.

Lemon Balm *(Melissa officinalis)*

Lemon balm supports headache relief, helps to encourage stress relief and restful sleep, and supports relief from menstrual cramps. Transplant seedlings outside, 18 inches apart, four weeks after your frost date. Lemon balm is a perennial that will come back every year in the same place. It requires rich, well-drained soil with partial shade from the afternoon sun.

Peppermint *(Mentha piperita)*

Peppermint promotes healthy digestive function and supports headache relief. Transplant seedlings outside, two feet apart, four weeks after your frost date. Peppermint is a perennial in zones 5 through 11. It is a fast-growing and spreading plant. Plant it where it won't be in the way, or plant it in a pot where it will be contained. It likes rich, moist soil and grows best in partial sun/shade.

Thyme *(Thymus vulgaris)*

Thyme is used to support healthy lungs and respiratory system, and corrects fungal imbalances. Transplant seedlings outside, 12 inches apart, two weeks after your frost date. Thyme is a perennial that will come back every year in the same place. It needs well-drained soil in a hot and sunny location. Allow the soil to dry out in between waterings.

Herbs in a Pollinator Garden

Having more pollinators visiting the garden will improve pollination, which increases the productivity of crops. These herbs will attract pollinators, and many of them have edible and medicinal uses, too. Whether you have time to grow food for yourself or not, providing food and shelter

for buzzing and fluttering pollinators is a way to be a good citizen. Their survival is threatened by environmental toxins and shrinking habitat.

Bee Balm *(Monarda didyma)*

Bee balm is a perennial flower with blooms of pinks and reds that are a favorite of hummingbirds and other pollinating insects looking for nectar. Lacewings—a beneficial insect—prefer the foliage as habitat for laying eggs. The scent of bee balm will repel pests when planted near the vegetable garden, or in an orchard or food forest. The leaves of bee balm make a mint-flavored tea. Space plants 18 to 24 inches apart in a sunny location. Soil should receive adequate moisture but be well-drained. It will grow two to four feet tall.

Borage *(Borago officinalis)*

Borage is a self-seeding annual. Bees flock to the star-shaped purple flower, and so do other beneficial insects that can't get enough of its nectar. Its foliage is a big draw as shelter during the growing season. To plant borage, sow seed in a garden bed that is clear of debris, then cover the seeds with ¼-inch of compost soil. Borage can be sown in the early spring or late fall, and if left to flower, will readily self-seed from year to year. It will grow two feet tall.

Comfrey *(Symphytum x uplandicum)*

Comfrey is a perennial herb with purple, bell-shaped flowers. It is a popular source of nectar for many pollinators and beneficial insects. The giant leaves provide habitat and an egg-laying site for many types of beneficial insects. Mulch the garden with the leaves to nourish the soil. See the entry earlier in this chapter for planting instructions and other uses. Comfrey grows four feet tall.

Coriander *(Coriandrum sativum)*

Coriander is an annual and a member of the carrot family, all members of which are favored by pollinators. Coriander attracts ladybugs, hoverflies, parasitic wasps, and lacewings, too—all beneficial insects. See above for planting instructions and for culinary uses. Coriander grows two feet tall.

Dill *(Anethum graveolens)*

Dill is another annual member of the carrot family, and as such, its clusters of yellow flowers form a landing pad for many types of beneficial insects such as ladybugs, hoverflies, parasitic wasps,

and lacewings. Dill will grow two to four feet tall in full sun or partial shade. See the entry earlier in this chapter for planting instructions and other uses.

Fennel *(Foeniculum vulgare)*

Fennel is a member of the carrot family and is similar in appearance to dill. Fennel has many of the same characteristics for attracting pollinators. Fennel has anise-flavored leaves and seeds. Dill and fennel should not be planted near one another since they will cross-pollinate. It is usually taller than dill. See the entry earlier in this chapter for planting instructions and more details about fennel's medicinal benefits.

Lovage *(Levisticum officinale)*

Lovage is a tall, attractive, perennial plant, reaching up to seven feet tall. The greenish-yellow flowers attract many kinds of pollinators, and beneficial insects seek shelter in the foliage. The stems are used like celery in cooking, the young leaves have a lemon-parsley flavor and are often used in salads, and the seeds are used for many of the same culinary uses as celery seed. Lovage grows best in full sun and rich, moist soil. Start from seed indoors six to eight weeks before your average last frost date, or space transplants 36 inches apart.

Roman chamomile *(Chamaemelum nobile)*

Roman chamomile is a low-growing perennial ground cover with small, daisy-like flowers. Pollinators of all kinds enjoy the button-like flowers, and parasitoid wasps—a small, beneficial insect for the vegetable garden—finds shelter in the foliage. The flowers make a soothing, medicinal tea. Mulch the garden with the spent plants at the end of the season to nourish the soil. It likes full sun and well-drained soil. Roman chamomile makes a nice ground cover in non-walkable areas. It is best planted by root cuttings 12 inches apart, and will grow to 12 inches tall.

Sunflowers *(Helianthus spp.)*

Sunflowers provide nectar for many kinds of pollinators and beneficial insects, and the giant leaves provide shelter. There are many kinds of sunflowers, from annuals to perennials, with a variety of flower head sizes. All are popular with pollinators and will grow anywhere from two feet to 16 feet tall. All the seeds are edible, but annual sunflowers with large seeds are more popular for this use. To plant annual sunflowers, sow seeds six inches apart in a full-sun area, after the

Bee balm

Comfrey in bloom

Yarrow

danger of frost has passed in the spring. Once established, thin the plants to two feet apart. Perennial sunflower seeds are best sown in the fall or by planting transplants two to three feet apart in a sunny or partially sunny area.

Yarrow *(Achillea millefolium)*

Yarrow is a perennial flower that by far attracts the largest variety of beneficial insects. Pollinators are also attracted to its lacy foliage for shelter. The leaves and flowers of yarrow have many medicinal uses. The scent of yarrow will repel pests when planted in the vegetable garden, orchard, or food forest. Mulch the garden with the spent plants at the end of the season to nourish the soil. Space plants two feet apart, and expect a height of two to four feet.

Create an Herb Spiral

An herb spiral is a compact, vertical garden design that makes it possible to grow a lot of herbs in a small space. I like to use this technique for growing culinary herbs outside the back door, so I can grab herbs quickly while cooking. Just a small handful of fresh sage leaves—for example—can really pack flavor into a soup or casserole. The convenience of a back-door herb spiral makes it more likely that I'll use the herbs, and considering the nutrient density of small amounts of herbs, this is good news!

Other people might like to use an herb spiral for growing a medicinal garden by the back door so that its first aid benefits are conveniently located when needed.

Coiled like a snail shell, an herb spiral winds up and around, with the center of the coil being the highest point of the growing bed. While an herb spiral can be of any size, it is generally a mound that is 18 inches to three feet tall at its central, top-most point, and about five to seven feet in diameter. The spiraled coil is usually outlined with rocks or bricks.

The height of the bed creates micro-ecosystems because there are sections facing in each direction, with some sections (south and west) receiving more sun than the north- and east-facing sections. Additionally, the lower sections will stay moister longer, while the tallest point will be the driest. This allows you to grow a large diversity of herbs. For example, rosemary, sage, and thyme will enjoy the sunny, dry sections, with herbs like basil, chives, cilantro, and parsley enjoying the partial shade and moisture of the northern, and lower sections. Oregano and marjoram are versatile and can work well wherever there is space.

In a medicinal herb spiral, lavender and thyme will grow nicely in the sunny, dry sections while comfrey and lemon balm will enjoy the moister, shadier sections. Dandelion and garlic

are more versatile and can be placed wherever there is space.

See "How to Build an Herb Spiral" in this chapter for building an herb spiral that is 18 inches tall and five feet in diameter.

When I first started gardening, herbs seemed like accessories that were nice to have, but not essential to the craft of growing fruits and vegetables. Now I know that seasoning food, growing my own medicine, and contributing to a healthy ecosystem are just as important. Herbs will always be an essential component of my garden, and if by chance I find myself too busy to grow vegetables, I will delight in still having my herbs.

Snail shell Photo by Sid Mosdel via Flickr

Mini herb spiral planted with cilantro,chives, sage, thyme, and rosemary

HOW TO BUILD AN HERB SPIRAL

Materials

- Hoe, hard rake, or flat shovel
- Cardboard to cover 20 square feet
- 25 cubic feet (about 1 yard) of any combination compost soil, worm castings, homemade compost, chemical-free aged manure, or aged leaf mold
- Bricks or rocks for border
- 48-pound bag leveling sand (optional)
- Stakes or flags
- String
- 2 (5-gallon) buckets of pea gravel
- Herb seedlings

Instructions

1. Locate the area where you would like to construct the herb spiral and measure a circle that is 5 feet in diameter to be sure it clears buildings and sidewalks.
2. Use the hoe or shovel to level the area, and remove any woody plants and vines.
3. Lay cardboard over the area, overlapping the ends by 6 inches to ensure that weeds don't creep through. Non-woody vegetation will be killed by the cardboard, so no need to remove it.
4. Locate the center again and carefully insert a stake through the cardboard. Tie string to the stake and measure a piece that extends 2.5 feet from the stake and cut the rest off.
5. Use the string as a guide to outline the perimeter of the circle with bricks, orienting the bottom of the spiral to face south.
6. Lay out the shape of the spiral with the bricks, keeping an even width of the growing area throughout (about one foot wide).
7. Locate the outer-most, beginning point of the spiral. From that point, lay a piece of string across the garden in a straight line, essentially drawing a diameter line that cuts the spiral in half along the north-south line. Use stakes or flags to mark points along this line.
8. Now it's time to use the bricks to gradually up the height of the border using the stakes as a guide, so that the middle point of the spiral is the highest point.
9. Build the wall two bricks high between the outermost points of the diameter. Following the spiral toward the center, build four bricks high between the points along the diameter. The next section along the spiral will be six bricks high. Finally, the last, topmost section of the spiral will be eight bricks high. Use leveling sand between the layers of bricks for a sturdier fit.
10. In the northern-most half of the spiral, add the pea gravel on top of the cardboard, making it gradually deeper to mirror the increasing height of the bricks. The pea gravel will help to stabilize the taller sections of wall, and it will also allow good drainage for the herbs that enjoy the highest and driest spots.
11. Now fill in the herb spiral (on top of the cardboard and pea gravel) with layers of organic soil material to match the height of the bricks. Water everything in or wait for a good rain to settle the contents of your herb spiral, then check the level of soil again and fill to match the height of the bricks.
12. Time to plant! Lay out the herbs, matching each herb to the conditions that it prefers (hot/dry, shady/moist). Be sure to give your plants proper spacing.

Your herb spiral will last a long time, be a low-maintenance treasure, and produce abundantly.

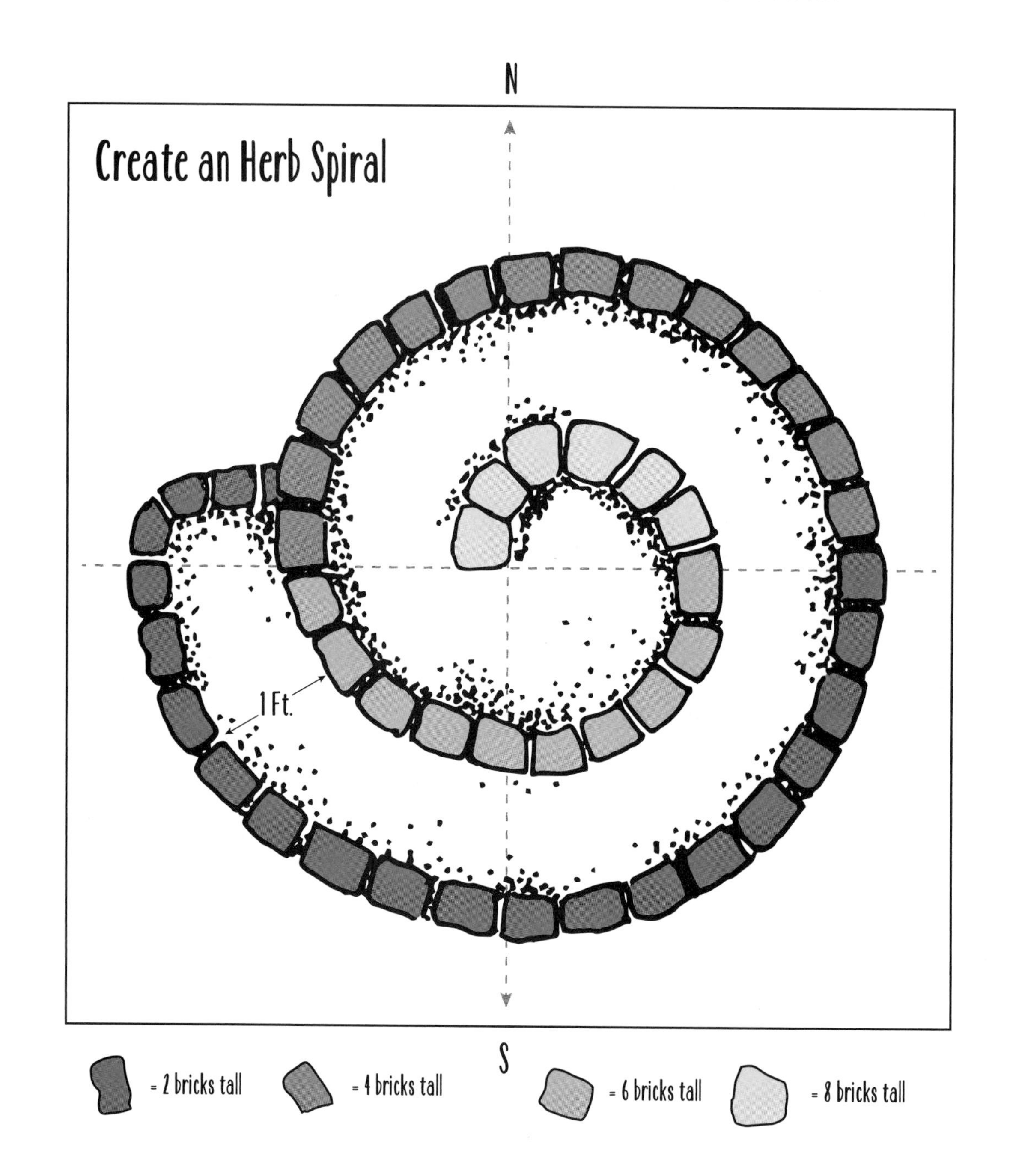
N
Create an Herb Spiral
1 Ft.
S
= 2 bricks tall
= 4 bricks tall
= 6 bricks tall
= 8 bricks tall

TO BE SUCCESSFUL, THE FIRST THING TO DO IS FALL IN LOVE WITH YOUR WORK.

SISTER MARY LAURETTA

PART 2:
BECOMING A MICRO-FARMER

7 ORGANIZING YOUR MICRO-FARM

IT TAKES AS MUCH ENERGY TO WISH AS IT DOES TO PLAN.

ELEANOR ROOSEVELT

Are visions of bountiful gardens dancing in your head? When it comes to having a successful suburban micro-farm, knowing how to plant is an important step, but it's only half the battle. The other half is streamlining how the garden is managed so you can get a harvest without neglecting your busy life.

I was sorely lacking in a streamlined approach my first few gardening seasons. I wasn't organized—I didn't keep records, so I didn't know when things should be, or had been, planted. I lacked a plan other than "I want a garden." When I got outside, I had no idea how to prioritize my time, and it made gardening chaotic and not very fun. I didn't feel a strong connection to my garden because my mind was caught up in the decision-making of what to work on in the moment.

I longed for a process to manage my micro-farm, and a central location in which to keep all of my gardening resources and tasks, eliminating the need for multiple file folders, random scraps of scribbled notes, and a slew of web pages bookmarked on my computer. After a few years of searching, I created my own **Micro-Farm Organization Process**, which has worked wonders for me. In this chapter, I share this process with you, as well as the materials that help me stay organized.

LUXURY IS FEELING UNRUSHED. IT IS DESIGNING A LIFE THAT ALLOWS YOU TO DO WHAT YOU WANT WITH HIGH LEVERAGE, WITH MANY OPTIONS, ALL WHILE FEELING UNRUSHED.

TIM FERRISS

These supplemental materials will help define your priorities so you can enjoy your time in the garden. The key to feeling relaxed and connected to your micro-farm is to reduce the mental energy needed when you're working in the garden. In essence, you want to reduce "decision fatigue," as entrepreneur Tim Ferriss calls it. According to his research, "The more decisions you rack up in one area, the fewer decisions you can effectively make elsewhere."

This means that the various real-life decisions you have to make in a given (busy) day, the less likely it is that you will be able to walk out to the garden in the evening, make decisions on the fly, and use your limited time wisely. Ferriss' recommendation is to use checklists to remove the mental energy of decision-making.

Enjoying the view of the garden from the front porch

The materials in this chapter will remove some of the stress of running a micro-farm. They will help you hone in on the top priorities so you can conquer the garden season confidently. I'll show you how to schedule around your busy life while still holding the garden as an "unrushed" priority. To download the supplemental materials in this chapter, go to www.TenthAcreFarm.com/tsmf-companion.

MICRO-FARM SUPPLEMENTAL MATERIALS

To help you get organized, I'm sharing with you the supplemental materials that helped me. To access them, along with video instructions, go to www.TenthAcreFarm.com/tsmf-companion. These materials are all in Google Docs and are editable once you make your own copy. They can also be printed in PDF format.

- Seed Starting & Planting Worksheet
- Purchase Recommendations
- Sample Monthly Checklists
- Monthly Calendars

Make a Garden Sketch

So far in the book, we've discussed things like how to build garden soil and suggestions for choosing what to grow based on your time availability and specific growing conditions. Now is the time to flesh out the actual organization of your vegetable garden area and what you will plant.

First, take measurements of your garden area. On graph paper and with a pencil, sketch your planting area, accommodating your actual measurements. For example, one square on the graph paper might equal one foot. Label north and south. Give each garden bed a name, such as Strawberry Bed, Bed #1, or East Garden. Label where pathways, lawn, play areas, and other areas exist around the garden. Don't yet label where specific crops will go.

When you're satisfied with your sketch, make a copy of it so you have a blank master sketch that you can copy each year. Alternatively, if you'd like a digital solution, there are many great garden planning programs online.

Now take a *copy* of your sketch and give it a title, such as "Backyard Garden 2016," so it will serve as a proper record going forward. You may even want to get a three-ring notebook in which to keep it and other garden records. Of course, if you are planning digitally, your files will be electronic. Try organizing your files or paper records by location ("West Garden 2016"), year ("Garden Plan 2016"), or crops ("Salad Garden 2016")—whatever works best for you.

Front yard garlic harvest

Choosing What to Plant

Now is the time to begin filling in your garden sketch with what you want to plant for the season. Gardeners can get all googly-eyed as they leaf through seed catalogs. There are so many beautiful and interesting vegetables out there, let's see if we can hone in on what will work best for you.

Why not start with what your household already eats? Make lists of what you like to eat and what you don't like or can't eat. Choose only from the "like" list while you're getting started. You can be more adventurous in the future as you become more comfortable with your garden process.

Some other considerations: Are you in a condo and planting a container garden? Don't forget to take advantage of vertical space. Or, perhaps you have to grow vegetables in poor soil? In our case, the garden beds are partially shaded by neighbors' trees. The shade provides a challenge, but it's also an opportunity for us to grow salads, leafy greens, and root vegetables almost year-round.

GARDENERS LEARN BY TROWEL AND ERROR.

AUTHOR UNKNOWN

If your garden is susceptible to a particular pest, choose pest resistant seed varieties. Hybrid seeds offer pest resistance or other qualities that allow you to grow a crop in a challenging situation. For example, container garden seed varieties have been developed to grow shorter plants with smaller root systems, and varieties of spinach have been developed that are slow to bolt in the heat of summer.

Saving seeds from crops that do exceptionally well in your garden will improve the success of your garden as time goes on. I've been able to save years' worth of herb and flower seeds such as calendula, chives, coriander, dill, fennel, garlic chives, and parsley, as well as bean, broccoli, collard green, and kale by growing heirloom seeds. See chapter 9 for more details about saving seeds.

We all deal with some kind of challenge. You might deal with pest problems, shade, or space limitations, but ultimately you must make some tough decisions about what to grow. By identifying what you love to eat and what the appropriate crops are for the conditions of your growing space, you can hone in on what's worth your time to reap a lovely harvest.

Don't forget, more recommendations can be found in chapter 4 for deciding what to plant, how much to plant, where to plant, and for dealing with challenging situations.

 Make Seed and Supply Purchases

Once you've designed your garden and decided what to grow, it's time to buy seeds, plants, gardening equipment, and supplies. This can be both exciting and a little nerve-wracking as you try to decipher essential purchases from discretionary ones. Let's wade through together and figure out how to be kind to the budget while having the essentials.

Buying Seeds and Plants

Buying seeds is one of the first things we think about when January rolls around. How do you choose where to buy your seeds? Seeds from the hardware store will work, but personally, if I'm going to go out of my way to grow my own food, then I want it to be healthy food: chemical-free with a commitment to the environment and future generations. Without this commitment, the food I grow from conventional seeds would be no healthier than that of the grocery store, making me question the time and effort of my actions.

The rules governing the growing of *seed* crops are much less stringent than those regulating the growing of *food* crops. This means that some chemicals—illegal to use on kale that I eat—are perfectly fine to use on plants that are grown for kale seeds.

Oftentimes the chemical residue remains and will contaminate my soil—even if in small, seed-size doses. I worry that it would concentrate over time from sowing seeds year after year. Though there haven't been any studies conducted on the topic, I'm not sure I want to find out that the soil I've worked so hard to build might be contaminated. Not only that, but the plants from those seeds have adapted to grow alongside chemical treatments. What happens when I try to grow conventional seeds using organic methods? I'm more likely to run into pest problems because the plant has adapted to growing with chemicals. This might not always be the case, but choosing the natural route gives me the most confidence that I know what I'm getting.

Choose a seed company that is committed to high quality, chemical-free seeds, fair pricing, and a diverse heirloom seed selection. A company that is geographically close to your region will have seeds that are well adapted to your climate.

> LEARN TO BE AN OBSERVER IN ALL SEASONS. EVERY SINGLE DAY, YOUR GARDEN HAS SOMETHING NEW AND WONDERFUL TO SHOW YOU.
>
> *AUTHOR UNKNOWN*

You might wish to look for companies that have pledged support to the Safe Seed Initiative, a collection of growers who are committed to propagating seeds without the use of genetic engineering. For a link to companies who have signed the Safe Seed Pledge, go to www.TenthAcreFarm.com/tsmf-companion.

Often, you can buy plants from the same company from which you buy seeds. In addition, seek out plant propagators in your local area that sell chemical-free plants. Farmers' markets and local gardening groups are good places to check, since regular nurseries tend to use chemical controls. The chemical residue is not only bad for us, but also pollinators.

Buying Garden Equipment and Supplies

When I first began gardening, I was totally overwhelmed with designing the proper garden, selecting what to grow, and learning how to grow it all. The one thing I didn't want to take time to research was which companies I should give my business to when it was time to buy supplies and equipment.

On a limited budget, I didn't want to buy a bunch of cheap stuff that wasn't going to stand the test of time. At the time, I couldn't find good reviews of garden products from people I trust. So I took a guess and bought what I thought I needed. Some stuff was super-useful and lasted a long time, some wasn't and didn't.

To save you the trouble, I created a shopping guide in which I review all of my favorite garden supplies and equipment that received good reviews from others and has stood the test of time. For products that didn't match up to my standards, I state what I would have bought instead.

If you'd like to check out my shopping guide, get the link at www.TenthAcreFarm.com/tsmf-companion.

Plan Your Seed-Starting and Planting Schedule

Figuring out what to plant when, especially when it comes to starting seeds indoors, can be a lot of work. To determine the date to start tomato seeds, for example, you will first identify your spring frost date and count backward on the calendar. Tomatoes are started about six weeks before the spring frost date. Counting backward on the calendar from my frost date, I can determine that I will start tomato seeds indoors around March 8.

Seeds from a community seed swap

A Spreadsheet for Your Thoughts

The first thing I did when I developed my **Micro-Farm Organization Process** was to create a spreadsheet that would help me organize my seed-starting and planting schedules.

To determine your start dates for seeds sown inside (under lights), seeds sown outside, and when to transplant seedlings, all you'll need to do is download my **seed starting & planting spreadsheet**, enter your region's spring frost date, and your planting dates will automatically be generated (Decision-making eliminated!). To download the spreadsheet and to find your region's first and last frost dates, go to www.TenthAcreFarm.com/tsmf-companion.

On the spreadsheet, I've also included information about the most popular vegetables and herbs, companion planting suggestions, and general information about planting fruit. Once you've downloaded it, the spreadsheet is completely customizable, and you can edit it to make it the most useful to you.

The spreadsheet will help you keep records for this gardening year and each one into the future, since you will always have access to download a new copy and edit it with the current year's specifications. It's the perfect activity to work on while you wait on the arrival of your seeds and gardening supplies!

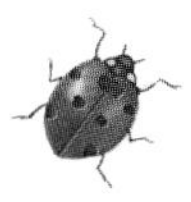

Month-by-Month Micro-Farming

The following is an overview of how I expect to spend my time throughout the months of the year, based on the rhythms of the seasons. This is a simple guideline, based on my experience of micro-farming in USDA hardiness zone 6a. Your experience may be different, depending on the climate and growing conditions of your area.

January

January is a restful time of passive harvests that reap the rewards of labor worked in the previous fall.

It's also a time of reflection and planning: looking to the past for growing wisdom that informs the future garden. I delight in browsing seed catalogs and cooking comfort meals that are chock-full of stored root vegetables and fresh, leafy greens. It's a time to tidy the farm both inside and out. Supplies and tools are cleaned and organized.

February

February is the month to wake from hibernation, collect seed-starting supplies, and begin the annual journey of seedling parenthood. Keeping a watchful eye on the young seedlings brings routine to the start of another season. Soil amendments can be added to beds now and cold frames can be planted with leafy greens.

In the kitchen, slow-cooker meals take advantage of fresh, leafy greens and preserved/stored vegetables.

March

March is when the gardening season increases in intensity and deadlines loom: Seeds are started indoors, in the cold frame, and outdoors, fruit trees and other edible perennials are planted and pruned, and procrastination for building new garden beds has reached its limit. Yet the crisp, fresh air holds a steady hopefulness and excitement for a bountiful garden season.

Compost soil is added to garden beds in the spring

April

We continue our careful watch over seedlings started inside, while the weather dictates windows of opportunity for sowing and transplanting outside. Glimpses of beautiful green life pops up in the form of garlic stalks, which handle the unpredictable spring weather with fortitude. This is a good month for infrastructure projects such as constructing compost bins or rain catchment systems. Time to put away the cold frame until fall.

IN THE SPRING, AT THE END OF THE DAY, YOU SHOULD SMELL LIKE DIRT.

MARGARET ATWOOD

May

The symbol for May is the busy bee! Although I usually stick to a 15-minutes-per-day gardening schedule, I try to sneak in extra minutes this month weeding and preparing garden beds, and transplanting seedlings. Sweet, juicy strawberries and fresh asparagus reward our toils.

The accessory gardens get some attention this month: window boxes and container gardens are planted, and potted perennials are repotted with new soil.

June

Flowers bloom, abundant rains keep everything green and perky, pests are nowhere in sight, and the harvests are rolling in. This month, we delight in being outdoors surrounded by beauty and abundance. Berry harvests come in the form of black raspberries, currants, strawberries, and cherries. We continue to sow seeds and plant seedlings. It's also a good time for mulching.

July

Steady harvests begin to direct more work to the kitchen for food preparation and preservation. This is the month to enjoy a few lazy summer days before the dog days of August and the heavy late season harvests set in! It's also time to design the fall garden.

A January harvest of Brussels sprouts, chard, collard greens, and daikon radish

An August harvest

A fall harvest

August

The hot and humid dog days of summer are here, but the garden persists. Overwhelming abundance has us carrying heavy harvests from garden to kitchen and counting our blessings. Harvests are transformed into delicious, fresh meals only experienced at this time of year. Surplus is converted into preservable forms. Now, we plant fall crops.

September

As the sun begins to wane, vegetable plants sense the impending end of their short lives and enter a last-ditch effort to produce progeny through seeds and fruits.

Wildlife scurries and buzzes about, eating well and storing food for migration or for the long, cold winter ahead. Fall radishes are sown and dead or diseased plant matter is removed from garden beds.

October

The garden seems to exhale in October, offering a kind of sadness in the finality, yet relief from the heat and the busy-ness of the season. Winter preparations include adding soil amendments and mulch to inactive gardens. The cold frame returns for fall and overwintering crops.

Next year's garlic goes into the ground and we head into the kitchen to put up tomatoes, beans, peppers, and more for wintertime eating.

November

The growing season is coming to a close, and gardens are being put to bed. We celebrate the orange and yellow colors of the season and bask in the golden light of the early sunsets. The last of the fruiting vegetables roll in now, while the leafy greens and root vegetables come into their own. It's a good time to plant fruit trees and berry bushes.

December

As the calendar year comes to a close, it can be as busy or as calm as we choose. It's a good time to build new garden beds so they can establish themselves over the winter. Fall crops are covered with row cover or cold frames to extend the harvest.

In the kitchen, we are busy making sweet treats and jams for the holiday season with frozen spring fruits.

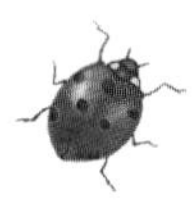

Checklists and Monthly Calendars

The **Sample Monthly Checklists** are what I use to keep track of garden tasks and when to do them throughout the year. The checklists help me visualize what's in store for any given month. They also help me remember once-a-year tasks that are easily forgotten, such as pruning the fruit trees or berry bushes. The use of checklists is Ferriss' number one recommendation for using time efficiently.

After I've decided what to plant and have completed my **Seed Starting & Planting spreadsheet**, then I fill in my monthly checklists with harvesting, planting, and sowing information.

Each monthly checklist starts with harvesting, then works in garden maintenance, sowing, and planting. Harvesting should be the number one priority. Even if you don't get the chance to plant anything else, at least you can reap what you've already taken the time to sow.

Once you download the checklists, you will be able to edit them and make them relevant to your farm, based on your climate, what you're growing, and what your priorities are.

Keep Notes: Observe and Reflect

One feature that I enjoy on the monthly checklists is the notes section. I almost always carry my monthly checklists to the garden with me (in my three-ring notebook) so I can cross items off as I complete them. The notes section is an excellent place to jot down important information—dates for when I noticed certain pests, when I planted certain crops, or when I added soil amendments, for example. I find that having the checklist in the garden encourages me to be more observant of my experience and to write things down as if I were a curious scientist:

> *"Kale leaves are especially chewed up today from cabbageworms, Sept. 1st. Note for next year: Research beneficial insects that prey on cabbageworms and what I should plant with kale to attract beneficials."*

MONTHLY CHECKLISTS

I prettied up the monthly checklists I use so that I could share them with you! Go to www.TenthAcreFarm.com/tsmf-companion to download and personalize them. These are SAMPLE checklists only, and include gardening tasks that are relevant to me based on regional growing conditions.

These notes and reflection will help you in subsequent years to become a better micro-farmer and more in tune with the challenges and features of your particular landscape. I also use it as a temporary harvest log until I get back to my computer. Remember that you can download **sample monthly checklists** at www.TenthAcreFarm.com/tsmf-companion.

FAQ: *WHAT IF I DON'T GET ALL THE TASKS DONE IN A MONTH?*

No worries, just add them to the list for the following month!

Monthly Calendars

So you've filled out your **Seed Starting & Planting Spreadsheet** and downloaded your **Monthly Checklists**. You're feeling pretty organized, but you're still not certain how to balance your busy "real life" stuff and all of this gardening stuff.

I'm a visual person, and using a monthly calendar helps me to see all of a month's tasks against my real-life schedule so I can make time for all of the seed starting, planting, and other garden activities. To get started, download and print the **monthly calendars** at www.TenthAcreFarm.com/tsmf-companion.

Add Seed Starting and Planting to the Monthly Calendars

I used to spend money each year on a wall calendar to write in all of my garden to-do items. I would take information from my **Seed Starting & Planting Spreadsheet** and write it on the calendar. For example, on March 8, I might write, "Start tomato seeds indoors," and on May 3, I might write, "Transplant tomatoes to the garden."

Rather than having to purchase a wall calendar each year—which didn't fit nicely in my three-ring garden notebook—I created a blank **monthly calendar template** to download and print. Fill in the dates for the current year, then transfer information from your seed-starting and planting spreadsheet over to the calendar.

Add Personal Life Requirements to the Monthly Calendars

When I write my real-life commitments on the calendar, such as work hours, "Grandma's Birthday Party," "Oscar's Karate Match," or "Family Vacation," I can see what days are available to schedule in garden tasks.

HOT CLIMATE GARDENERS

I've attempted to make the planting information in this book as universal as possible. The majority of gardeners will calculate planting dates based on spring and fall frost dates. However, it may be difficult for many hot-climate gardeners to achieve good germination rates on summer-sown seeds for the fall growing season. In general, hot-climate gardeners simply delay sowing seeds and planting fall seedlings until temperatures drop below 90 degrees F. Check with your local cooperative extension office (see www.TenthAcreFarm.com/tsmf-companion for appropriate planting dates, which may not follow your local frost dates closely).

Prioritize Tasks from the Checklists on the Monthly Calendars

The following garden planning tasks are listed on the January checklist:

- Sketch garden layout
- Decide what to plant
- Figure how much to plant
- Make seed and supply purchases
- Download and complete **Seed Starting & Planting Spreadsheets**
- Download and print **Monthly Checklists**
- Download and fill out **Monthly Calendars**
- Set up an indoor seed-starting system

March Checklist

Tasks to Complete	Date Completed	Notes / Observations / etc.
Harvest from under Protection		
beets, chard, spinach		
collards, kale		
leek		
lettuce		
Garden Maintenance		
Build garden infrastructure: rain catchment systems, fences, compost systems, garden beds, etc.		

When I schedule these garden tasks on the calendar (to complete during my 15-minutes-per-day) as though they were required events, I'm more likely to get them done. This helps me keep the garden as a priority, and it also means that I'm more likely to use my time efficiently when I work on gardening stuff.

That's all there is to the Micro-Farm Organization Process! Because all of these resources are available to you at www.TenthAcreFarm.com/tsmf-companion anytime, you are welcome to use them however you'd like. Although I have had the best success using all of the resources together, you can mix and match—or omit any that don't seem useful to you.

As you can see, there are a lot of things to think about when planning a garden and successfully getting it underway in the springtime. Avoiding decision fatigue will be essential to success for the busy micro-farmer, and using the supplemental materials mentioned in this chapter will help.

SEED STARTING & PLANTING

IT IS ONLY THE FARMER WHO FAITHFULLY PLANTS SEEDS IN THE SPRING, WHO REAPS A HARVEST IN THE AUTUMN.

B. C. FORBES

You've planned your garden to the hilt, chosen your crops, and scheduled tasks on your calendar. Now it's time to figure out how to grow everything you want to grow! In my first couple of years of learning to garden, I didn't start my own seeds indoors. Rather, I planted vegetable seedlings that I purchased at my local farmers' market. By my third year of gardening, I felt confident enough to begin learning how to start my own seeds indoors. Remember, you don't have to learn everything at once! In this chapter I'll share how I start my seeds indoors, how to sow seeds directly outdoors, and how to transplant seedlings into the garden for the best success.

Starting Seeds Indoors: A Step-by-Step Guide

Starting seeds indoors is quite the process to get accustomed to, but as with all things, it gets easier with practice. There's an initial investment to acquire the equipment and materials, so if you're new to gardening or are super-busy, it might be better to skip this step and buy your seedlings locally. If you go that route, be sure to ask the grower about his/her growing practices so you can choose chemical-free plants. After you make the initial investment of buying your seed-starting equipment, starting your own seeds will be the better economical choice.

I've developed my own system over the years, and you'll find that no two gardeners start seeds indoors exactly the same way. The following system has worked for the amount of growing that I do on my micro-farm. In fact, this system allowed my community garden team to grow hundreds of seedlings for our annual plant sale.

TIP: SEED STARTING EQUIPMENT

To see my purchase recommendations for seed starting equipment and materials, go to www.TenthAcreFarm.com/tsmf-companion.

Shelves for Seed Starting

Shelves are the first things you'll need for seed starting. They determine a lot of things, such as the size of your light ballasts and how much of the other equipment you'll need. I use the wall-mounted wire shelves in my mudroom—which I now call the "seed" room.

Free-standing wire shelving unit

Wall-mounted shelves with aluminum foil on the wall to reflect light

Seedlings under lights

You may prefer to buy a free-standing wire shelving unit or build your own wooden shelf unit. I think it's easiest if you can access both sides of the shelves, but I've managed fine with single-side access to my wall-mounted shelves.

Other Seed-Starting Equipment and Supplies

- Aluminum foil or Mylar reflective film for low light areas
- Fluorescent or LED grow light fixtures (2 per shelf—find the length appropriate for your shelving unit)
- Full spectrum fluorescent or LED light bulbs (2 bulbs per light fixture)
- Screw hooks for wooden shelves OR carabiners for wire shelves (2 per light fixture)
- Mountable fan (one for each shelf)
- Seedling heat mat (at least one per shelf, up to four per shelf)
- Thermometer/humidity monitor
- Programmable power strip
- Extension cords
- Waterproof table OR shop-style bench (one that can get wet and dirty)
- Standard nursery drainage trays (no holes—22"x11")
- Plastic seed-starter cell packs
- Seeds
- Seed-starting soilless planting medium
- Galvanized steel tub
- Plant markers
- Sharpie pen
- Watering can
- Garden gloves
- Pencil (sharpened)
- Plastic wrap (Saran-style) OR plastic domes to fit nursery trays
- Potting soil
- Plastic pots
- Cinnamon spice shaker (optional)
- Fish and seaweed fertilizer
- Garden scissors

Seed-Starting Instructions

These are the steps I take in the late winter/early spring each year to start my own seedlings.

1. **Assemble the shelves.** Assemble the shelving unit and set it in a level spot, in a room with as much ambient light as possible. My seed-starting room is in the basement with no ambient light, so I cover the walls with aluminum foil (shiny side out) to reflect the light. Or try Mylar reflective film, the indoor growing industry's solution for light reflection.
2. **Hang the lights.** Hang the lights above each shelf using screw hooks or carabiners, and mount a fan at the end of each shelf.
3. **Manage the temperature.** Set out the heat mats on each shelf, and set out the thermometer on the shelving unit or close to it.
4. **Manage the power source.** Mount the power strip and organize the extension cords so that all light fixtures are plugged into the timed outlets, and all the fans and heat mats are plugged into the always-on outlets. No need to turn anything on yet.
5. **Arrange the power cords.** Organize the power cords so they're out of the way as much as possible, and away from possible water spillages when the seedlings get watered. I like to use twisty-ties to reign in gangly cords.
6. **Assemble materials.** Assemble your trays, cell packs, seeds, seed-starting medium, steel tub, plant markers, Sharpie pen, watering can (filled), garden gloves, and pencil near your work area—table or work bench.
7. **Prepare the planting medium.** Fill the tub with seed starting medium and add water, one gallon at a time, mixing just until the "soil" clumps together. It should feel like a dry sponge. If water can be squeezed from it then it's too wet. To remedy, add more soil. Excess planting medium can be saved. When you're finished starting seeds, just let the medium dry out before storing it.
8. **Prepare the cell packs.** Take a cell pack and fill each cell to the top loosely with the soilless medium. Pack it in firmly, so that now each cell is only one-third to one-half full. Now fill each cell pack loosely with soil again to the top. This time, press lightly rather than packing down firmly.
9. **Choose a seed packet to begin.** Select a seed packet to work with and consult your garden plans to figure out how many plants of that variety to start. Always start a few extra just in case.
10. **Plant seeds.** Seeds should be planted twice as deep as their size. Plant at least two seeds per cell. Use the pencil to push the seeds in, then lightly press the soil on top so that the seeds are covered. Really tiny seeds, such as lettuce seeds, can just be pressed into the top of the soil mix rather than buried.

11. **Label the cell packs.** Label each cell pack with a plant marker.
12. **Fill drainage trays with cell packs.** Place each full and seeded cell pack in a drainage tray.
13. **Water seeds in.** When the drainage tray is full of cell packs, water each cell very lightly to encourage the seeds to make contact with the soil.
14. **Cover the trays for germination.** Cover the tray with plastic wrap or a plastic dome, and set it on a heat mat.
15. **Start the rest of your seeds.** Continue the process until all seeds have been started and all trays have been set on a heat mat.
16. **Turn on heat mats.** Make sure all heat mats are on. Lights and fans are OFF until the seeds have germinated.
17. **Check daily for germination.** Each day check for germination, briefly lifting the plastic wrap to allow some condensation to escape. Once seeds have sprouted, immediately remove the plastic covering.
18. **Monitor the temperature.** Check the thermometer. An optimal ambient temperature for seed starting is around 67 degrees F, though temperatures up to 73 degrees F are perfectly fine. I've used a space heater in addition to heat mats in my seed-starting room during cold winters. If you're short on heat mats, prioritize their use for the following heat-loving crops: eggplants, peppers, okra, watermelon, and tomatoes.
19. **Set the lights on a timer.** Once the seeds have germinated, turn the lights on and the fans on low. Turn OFF the heat mats for all but the most heat-loving plants listed in step 18, unless the temperature in the room regularly dips below 67 degrees F. Set the timers for the lights on the power strip: If your seed starting room has ambient light from nearby windows, it's ideal to allow your plants to follow the natural cycle of daylight hours. Program your lights to turn on at sunrise and turn off at sunset on the longest day of the year in your region (about the time of the summer solstice—June 21 in the northern hemisphere). For example, in my region I would turn the lights on at 6:00 am and off at 9:00 pm, for a total of 15 hours of light. This keeps my plants on a natural cycle. You can check for the longest day in your area by going to www.TenthAcreFarm.com/tsmf-companion. If you're starting seeds in a windowless room like me—with little to no ambient light—it will benefit your plants to keep the lights on 24 hours a day. In this case, the timer isn't necessary.
20. **Position the lights.** Using the adjustable chains on the light fixtures, position the lights 10 inches above the seedlings. Adjust the height of the lights regularly as the seedlings grow.

SEED STARTING FAQS

What if all the seeds in a tray don't germinate at the same time? Should I still remove the cover?

Try to group seeds in a tray that germinate at similar times. If that isn't possible, remove the cover when the first seeds germinate and put plastic wrap over the side that hasn't germinated yet.

I will be starting my seeds in the basement, where the ambient temperature is regularly below 67 degrees. Should I keep the heat mats turned on after germination?

A large basement would likely be too costly to heat with a space heater to increase the ambient temperature, so continue to use the heat mats, prioritizing them for the heat-loving vegetables. Most other vegetables will grow fine in the cooler temperatures. Some will prefer the cooler temperatures while others will simply grow a bit slower. Continue to monitor the heat mats closely and remove them when the room has reached the optimal temperature.

Oops, I've missed the seed starting date!

If you've missed seed starting for a particular crop, have no fear. Seedlings or young plants are often available from online or local nurseries and farmers' markets.

How should I store extra seeds once I've planted what I need for the season?

The best storage for seeds is a dark, dry, cool, airtight place. See chapter 9 for more tips on seed storage.

Caring for Seedlings

Brief daily visits to your seed starting operation can ensure that it is operating smoothly and that the seedlings have everything they need. Watering, thinning, fertilizing, potting up, and transplanting are some of the things that will need to be done to grow your seedlings and successfully transfer them to their outdoor home.

Cell packs are seeded and labeled

Watering the Seedlings

Once your seeds have germinated and are under the lights and fan, check frequently for dry soil (soilless medium). Let the soil dry completely before adding water, but don't wait so long that the seedlings droop. This will reduce the

chance of the seedlings getting a fungal disease called "damping off," which is a common disease in seedlings with too much moisture and not enough ventilation.

TIP: WATERING SEEDLINGS

Remove one cell pack from a tray and fill the drainage tray halfway with water using the watering can. Replace the cell pack. Set a timer for 10 minute intervals and check all the cells for moisture. When the cells are all sufficiently moist, carefully dump out the excess water from the drainage pan.

Watering from below instead of above reduces the incidence of damping off. If you have a damping off outbreak, immediately remove the affected plants and put them in the garbage. Sprinkle cinnamon on the soil of the remaining healthy seedlings to minimize the spreading of the damping off fungus.

Thinning the Seedlings

The seedlings must be thinned once they've produced their second set of leaves (first set of true leaves). If there's more than one seedling in a cell, remove all but the strongest and most vigorous seedling. Use the garden scissors to cut off the weaker and more slow-growing seedlings at the soil level. Don't pull them out, as doing so may disturb the delicate root system of the remaining seedling.

Fertilizing the Seedlings

The first leaves of a seedling get nutrients from the seed itself, requiring no fertilization on your part. As the seedling grows, however, it will develop its first set of "true" leaves. These true leaves signal that the seedling has used up its stores of nutrients from the seed and will now require outside fertilization.

Since the soilless medium is free of nutrients, you'll need to fertilize. To water the seedlings with fertilizer, add one tablespoon of fish fertilizer per one gallon of water and mix well, or follow the directions on the label. Use as part of the regular watering schedule.

Potting up the Seedlings

As the seedlings grow, you may need to transplant them out of their cells and into their own plastic pots. If your seedlings have more than two sets of true leaves, check your planting schedule. How long until they are planted outside? If there's more than one week until they will be transplanted outside, you'll want to pot them using potting soil (not the seed-starting medium). Keep in mind that the pots take up more space and might not all fit under the lights. Try to keep the pots in rotation between the lights and nearby sunny windowsills.

Transplanting your seedlings into their new home outside is a delicate process. See details under "How to Transplant Seedlings Outdoors."

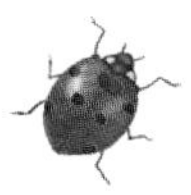

How to Sow Seeds Directly Outdoors

Many common vegetables are sown directly into the garden rather than indoors under lights. Beans, peas, and root vegetables are commonly direct sown. In addition, many people prefer to direct sow leafy greens, cucumbers, and squashes, rather than start them indoors. For planting instructions specific to each crop, see the Guide to Planting and Growing Vegetables in chapter 4.

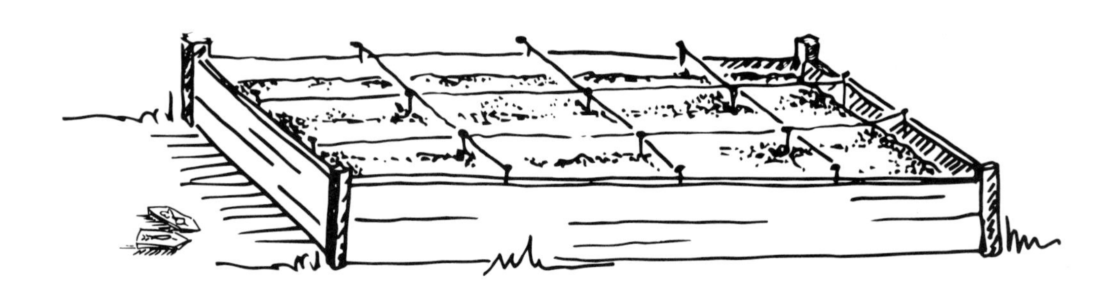

Prepare the Soil

Loosen the soil in your garden bed about six to eight inches deep using a digging fork. Ideally this will be done two to three weeks before planting. It's a good time to mix in soil amendments such as worm castings, aged manure, or compost, to create a rich environment for your crop.

Choose a Planting Day

Seeds are temperamental. They won't like soil that is completely waterlogged or soil that is completely bone-dry. Choose a day when the soil is slightly moist, and when there will be a few days following the planting without heavy rain showers (which will wash seeds away).

Mark the Rows or Square-Foot Grid

Use the edge of a trowel or garden hoe to mark a planting row. For example, to plant in a raised bed three feet wide by six feet long, create furrows to mark three long planting rows. Alternatively, to use the square foot gardening method, you may choose to mark off a grid of 18 square-foot blocks.

Sow Seeds

Sow seeds in the depressions marking the rows or broadcast the seeds in the desired square-foot block.

Cover and Pat the Seeds

If the seeds are really small, such as carrots or lettuce, cover them with about ¼ inch of soil. Medium-sized seeds, such as kale or beets, should be covered with ½ inch of soil. And finally, large seeds, such as beans or squash, should be covered with about one inch of soil.

Once the seeds are sufficiently covered, use your hand or the back of a hoe to pat down the soil firmly.

Label

Use plant markers to label the row or square so you remember what was planted where. I always think I will remember, but even with a clearly mapped out garden plan, I sometimes have trouble, especially when something hasn't germinated.

Water

Using a gentle spray, water the seeds in well. Seeds need to be watched closely for good germination. The soil can never dry out before seeds have germinated. Check daily and water with a gentle spray if needed. Once your seedlings have grown about six inches high, mulch with chemical-free straw or shredded leaves to help retain moisture and keep weeds down.

Start Succession Planting

Even the most experienced gardeners don't get 100% germination. You'll be able to fill in the gaps by sowing seeds regularly using the succession planting method. Check the succession planting schedules for each individual crop in the Guide to Planting and Growing Vegetables in chapter 4, or download the Seed Starting & Planting Worksheet at www.TenthAcreFarm.com/tsmf-companion. Most seeds are sown every 10 to 21 days, depending on the crop.

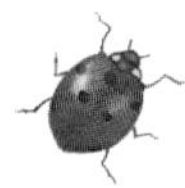

How to Transplant Seedlings Outdoors

Once seedlings reach a certain size—usually when they've developed their second set of true leaves—they're ready to be transplanted into the garden. Most seedlings will have some degree of sensitivity to frost. Check vegetables by their hardiness to frost in chapter 4.

The transplanting process can be stressful to seedlings. Whether you've grown your seedlings indoors or purchased them from a local nursery, following this simple procedure will make their transition easier.

Prepare the Soil

Loosen the soil in your garden bed about six to eight inches deep using a digging fork. Ideally, this will be done two to three weeks before planting. Mix in soil amendments such as worm castings, aged manure, or compost, to create a rich environment for your crop.

Harden them Off

Hardening off is the process by which seedlings become acclimated to the outdoor environment. The process starts one to two weeks before the seedlings are schedule to be transplanted outside. Sun, wind, hard rain, and temperature fluctuations can shock them. During the hardening off process, stop fertilizing them, but keep them well-watered.

Start by setting your seedlings outside in the shade for an afternoon on a dry, calm day when temperatures are in the 60s or higher. If extreme temperatures or heavy rain is predicted, it will be better to delay the hardening off process. Bring the seedlings indoors overnight.

The following day, expose your seedlings to a few hours of partial sun, back to the shade for the rest of the day, then indoors overnight. Each day give the seedlings more partial sun until the last few days when the seedlings can remain outside all day and overnight in a protected location with temperatures in the 60s.

Choose a Planting Day

Once the hardening off process is complete, choose a cloudy day with temperatures in the 60s or above. Moist soil that isn't waterlogged or bone-dry is ideal. If a heavy rain has recently occurred, it may be better to wait a few days for the soil to dry out a bit. In addition, check the weather forecast and choose a day when there will be a few days after transplanting without heavy rainfall. Light rains are okay, even ideal. Make sure seedlings are well watered in their pots.

Mark the Planting Location

Fill out plant markers for each seedling. Mark the planting spot for each new seedling with the plant marker. A tape measure may help you mark the distance between each plant.

Dig a Hole

Dig a hole for each seedling wider and deeper than the container, unless your seedling is a tomato, in which case double the planting depth. Turn the pot upside down and firmly tap the pot against the palm of your hand to free the seedling from the pot. Push the seedling out of the container from the bottom if necessary. Never pull on the stem of the seedling to remove it. Now, place the seedling in the hole so that the root layer is covered with at least a ¼ inch of soil. Gently press the soil down around the plant.

Label

Relocate the plant marker so it's visible in front of the plant.

Water

Using a gentle spray, water the seedlings in well, using a diluted fish fertilizer if desired. They will need quite a bit of water. Check the soil daily, it should not dry out around seedlings for at least a week while they adapt to their new home. Once all seedlings have been planted, spread a layer of mulch in between plants (but not touching the stems) to retain moisture and keep weeds down.

Baby lettuce ready for planting.

Plant markers are functional decorations in the garden.

The planting rows are marked with a garden knife, then filled with worm castings prior to planting.

Maintain Seedlings

Keep an eye out for wilting due to hot sun, and provide water or shade until the seedlings are established. Also watch out for chilling winds or frost and cover your seedlings to protect them. Row cover, buckets, plastic sheets, and cold frames are just a few implements that can help to protect seedlings from cold spells.

When you get your plants off to a good start, there's a good chance you'll produce something edible. Soon, you'll be producing all kinds of crops!

MAINTAINING THE GARDEN & HARVESTING THE PRODUCE

THERE CAN BE NO OTHER OCCUPATION LIKE GARDENING IN WHICH, IF YOU WERE TO CREEP UP BEHIND SOMEONE AT THEIR WORK, YOU WOULD FIND THEM SMILING.

MIRABEL OSLER

After the seeds and plants have been planted for the season, it's time to observe them during your daily, 15-minute visits. Any number of things can happen in between the time of planting and harvest that can affect the success of your garden. Adjustments will invariably need to be made, whether through mulching, watering, or wise pest prevention strategies. Keeping good records is key for a successful harvest.

But the fun part begins when you start harvesting your own goodies. There's nothing more delicious than the fruits and vegetables you grew yourself! What comes next—saving and storing seeds—is also gratifying. It's such a good feeling to save your own seeds that will be better adapted to the conditions of your micro-farm with each subsequent generation.

In this chapter, we'll take a look at some ways to manage these tasks. I'll share some tips for harvesting your crops at the peak of ripeness, then post-harvest, I'll address preparing the garden for winter and saving and storing seeds, so you'll be set up for an even better harvest next year.

 Daily Garden Visits

This is a good time for me to remind you how important it is to check on the garden daily—preferably with your morning coffee or evening happy hour drink for a more relaxed visit. In my first few years of gardening, I thought a daily garden visit meant a daily work session, so I avoided it, and received diminishing returns for my non-efforts.

However, a quick visit can allow you to take some mental notes (bare soil here—note to mulch; dry soil there—note to water; I spot a pest! What do I do?; and oooh, there's a zucchini ready for harvest that I would have surely missed if I had waited until Saturday!). When I discover these things during my morning coffee visit, I'm prepared to spend 5 or 10 minutes taking care of them at the end of the work day.

 Mulching in the Garden

Mulch is an essential component of a healthy garden. It is a biodegradable layer of organic material added on top of the soil in a garden or landscaping. In the permaculture garden, mulch mimics the forest floor, which is usually covered with ground covers, herbs, sticks, twigs, and leaf litter.

Mulch retains moisture, prevents weeds from germinating, reduces soil erosion, creates humus, fertilizes, and makes an attractive top dressing. Bare soil is damaged soil.

Shredded leaves and chemical-free straw are two common choices that will perform wonderfully. However, there are quite a few other types of mulch such as living mulch, green mulch, woodchip mulch, or grass mulch. Let's take a look at these.

Living Mulch

Living plants—either annual or perennial—can be used as mulch. The mulch plant will be planted around the major crop to shade and protect the soil, reduce weeds, retain moisture, and provide habitat for beneficial insects.

- Use annual living mulches in the vegetable garden. Examples of annual mulches are borage, calendula, nasturtium, and sweet alyssum, which can be seeded in between the rows.
- Use perennial mulches around perennial crops such as fruit trees, since their location is fixed. Examples of perennial mulches are comfrey, rhubarb, thyme, oregano, and white clover.

Living mulches are ideal for wet climates because their roots help absorb extra moisture. They can also be overwintered in garden beds to prevent soil erosion.

Nasturtium is a beautiful annual flower that can be used as living mulch and for pest deterrence in the vegetable garden.

MAKE THE MOST OF YOUR MULCH

Mulch will be used in different climates in different ways. In dry places, you can never have enough mulch! The more mulch, the better, to retain the moisture and protect the soil from the hot sun. Mulch produces a cooling effect, allowing you to have a more productive garden.

In rainy, humid climates, a light layer of mulch will keep the topsoil from washing away during rain events, but too much mulch can reduce air circulation and encourage disease. Since mulch has a cooling effect, deep mulch would have a negative effect in cooler climates, especially when growing heat-loving vegetables such as tomatoes or peppers. In my climate, where we have cool, wet springs and hot dry summers, I often wait until June to mulch, after the spring rains have passed.

Chard and Sweet Alyssum

One combination that I particularly enjoy is chard planted with sweet alyssum as living mulch. Together, they make a beautiful edible landscaping border along my front sidewalk. Not only is it beautiful, but the sweet alyssum also helps to combat slugs and flea beetles—pests that often attack the chard plant—by attracting beneficial insects. Mini combinations like chard with living mulches such as sweet alyssum will reduce garden maintenance while providing a beautiful and productive landscape.

Green Mulch

Green mulch is also called chop-and-drop mulch. Green grass, weeds, or other plants can be cultivated specifically for the benefits of protecting bare soil and providing nutrients (fertilizer) to major crops. Have you ever wondered why many of those pesky weeds have such deep taproots? Dandelion roots, for example, heal bare earth by plunging deep into the earth in search of nutrients. The "weed" then dredges up those nutrients in its leaves and roots. Take advantage of this free and abundant resource by mulching and fertilizing with dandelion leaves.

Sweet alyssum is used as an attractive living mulch between red chard plants in this edible border.

Broccoli mulched with weeds

Chive cuttings make great mulch around fruit because they mask the sweet smell and deter pests.

To use weeds as a green mulch, cut them at the soil line and lay the green leaves (not seed heads) on top of the soil beneath—but not touching—the garden crops. For a more attractive look, green mulch can be topped with a layer of leaf mulch or straw.

COMMON GREEN MULCHES THAT FERTILIZE

- Chickweed
- Chives
- Comfrey
- Dandelion
- Lamb's quarters
- Parsley
- Purslane
- Rhubarb leaves
- Yarrow

Comfrey

Comfrey is an herbaceous plant with beautiful purple flowers. It's also rich in nutrients. The large leaves of the comfrey plant are fast growing, and once cut, can grow back to a large size within weeks. This cutting doesn't hurt the plant. To use comfrey as a nutrient-rich, green mulch, plant comfrey near the vegetable garden or under fruit trees, and chop-and-drop its leaves five or six times throughout the growing season to spread around the garden. To read more about comfrey, see chapters 6 and 11.

Brown Mulch

Brown mulches such as leaves, straw, or wood chips are generally abundant in the suburbs. Here are some ways to take advantage of these free or low-cost resources.

Leaf Mulch

Leaf mulch is an attractive top-dressing in the garden that also helps retain moisture. We collect leaves from many of our neighbors and pile them in wire bins for use throughout the year. Use a lawn mower or leaf mulcher to shred the leaves prior to storing or using them in the garden.

Most types of leaves will work wonders for the garden; however, walnut leaves should not be used, as they have a chemical called juglone that suppresses healthy plant growth in all but a few plants. Oak leaves should be used only when mixed with other leaves because they break down slowly and contain tannins that can affect the soil composition.

Alliums (garlic, onions, chives, etc.) deter pests. Here dried garlic leaves are used as brown mulch under collard greens.

Wood chip mulch is used under fruit trees.

Leaf mold (leaves that have composted for two or three years) is an excellent soil conditioner. It is an essential component of our soil mixture, which reduces the amount of compost soil we need to purchase when developing new gardens.

Wood Chip Mulch

Wood chips are often delivered for free or a minimal charge from a local tree service. They are great to use in pathways or as mulch around perennials. Never let wood chips make contact with stems or trunks of plants.

Wood chips that have broken down for three years or more are a gold mine of a soil conditioner. Use the composted wood chips without caution in the vegetable garden, under fruit trees as an attractive top dressing, or in the grass mulch combo in place of leaves.

Grass Mulch Combo

A study at Michigan State University in the 1990s researched leaf mulch and how it contributed to weed suppression and fertilization. They found that leaf mulch did not serve much value as a weed suppressor or fertilizer source by itself.

When coupled with a nitrogen source (like green mulch), the weed suppression and fertilization levels went up. So I'm a proponent of composting in place just like you would in your compost bin, by layering the greens and the browns—the nitrogen and the carbon layers. Many gardeners have expressed concern that this practice ties up nitrogen in the soil and diminishes garden output, but this is only true if the material is mixed into the soil. As mulch, it will act as a slow-release fertilizer and weed suppressor.

Though I use many types of green mulch around the garden, my favorite combination is using the abundant grass clippings from my neighbor. I pile about one inch of green mulch in the garden beds (not touching the plants) and top them with two to three inches of shredded leaves or other brown mulch. This method has allowed me to keep more organic material out of the waste stream. It has also solved my problem of not having enough space in our compost bins to properly break down all the grass clippings from the neighbors.

Mulching the Garden for Winter

Another mulch option is using livestock manure (horse, cow, chicken, rabbit, etc.) as your green mulch/nitrogen source. In the fall, spread a two-inch-thick layer of manure over the beds and top with a layer of shredded leaves, chemical-free straw, or composted wood chips.

Grass clippings are topped with shredded leaf mulch in this bed of kale interplanted with onions to reduce weeds and watering.

All in all, mulching is an important component of lasting soil health, water conservation, and time savings for the micro-farmer. When the soil is happy, the plants will be happy, too. You'll experience fewer pest problems and regenerate your soil's fertility at the same time.

 Watering

Keeping the garden sufficiently watered is key to a successful harvest. Hand irrigating, unfortunately, can be time-consuming, so I believe it should be a last resort. You can reduce the need for hand watering through mulching and by using rain harvesting techniques such as swales and rain gardens (see chapter 11). These are good strategies to use because not only do they reduce the amount of chlorinated water used in the garden, but they also save time.

However, in dry climates and hot summers, watering the garden by hand will likely be a necessity at some point. Let's look at some tips for watering as efficiently and effectively as possible.

Have you ever had trouble keeping seeds watered sufficiently until they have germinated? Me, too, until I discovered the seed blanket—also called floating row cover or garden fabric—which can be laid over seeds just until they germinate to help hold in moisture and to keep them from washing away in the rain. Look for the summer weight fabric that transmits at least 85% of sunlight. Be sure to remove the fabric once the seeds have sprouted.

According to John Jeavons, author of *How to Grow More Vegetables*, the ideal time to water the garden is in the evening, two hours before sunset during summer. This is because plants do most of their growing overnight. If watering occurs in the morning, by the time the plants are ready to grow, much of the water has already evaporated.

However, a morning watering is more beneficial than a midday watering due to evaporation from the hot midday sun, and ANY watering is better than none! In other words, don't let your schedule keep you from giving your plants a good watering, no matter the time of day. And did you know that more water will evaporate on a windy day than on a sunny, still day?

Water the Soil

When you water the garden, you are actually watering the soil. Like a sponge, the soil in turn waters the plant roots. Moist soil is a good environment for beneficial soil microbes, which help plants process nutrients and aid in essential chemical and metabolic functions of the plant. Soil microbes help minimize the amount of water your garden needs to survive. Attract more of these beneficial soil microorganisms to the garden by enriching the soil with organic matter. See chapter 3 for more ideas on improving the soil ecology.

The Gentle Spray Method

A hard rain during drought will hit the soil like pavement and run off, taking topsoil with it. When watering, adjust your nozzle spray to mimic a soft rain, turning the nozzle upward if possible. This will ensure that the majority of the water is soaking in and isn't compacting soil or creating soil erosion. Check to see if the soil moves when watering. Standing farther away will allow you to mimic soft rain while watering a larger area at once.

Special Watering Needs

There is no one-size-fits-all strategy for watering. Each type of plant and soil will have its own preferences. For example, plants in the cabbage/brassica family (broccoli, kale, Brussels sprouts, etc.) love wet leaves, so they like to be watered from above. On the other hand, tomatoes,

legumes, squashes, and melons are more susceptible to mildew when the leaves are wet, so they like the spray to be directed to the soil next to the plants, avoiding the leaves as much as possible.

While sprinklers are one way to avoid hand watering, they are largely inefficient because a high percentage of the water will evaporate into the air. Drip irrigation, on the other hand, is a preferred method that irrigates at the soil level, allowing water to be absorbed into the ground. It also keeps water away from the leaves of plants, reducing the susceptibility of disease.

During a dry spell, daily watering works well for delicate annual vegetables. You may be tempted to skip a day if it's rained. Check for moisture in the top two inches of soil, and only skip a day if they are moist. Remember, some veggies—like tomatoes and peppers—can have roots 10 inches deep, and a single light rain shower during a drought may not offer sufficient moisture.

Fruit and other perennials aren't so keen on daily waterings. Because they have more established, permanent root systems, they prefer a deep, weekly watering. Be sure to mulch your fruit trees, bushes, and strawberries well.

Clay soil is a common challenge that many gardeners deal with. When it comes to watering, clay soil responds better to more frequent, lighter waterings than to occasional, heavier waterings. Thus, in drought, watering every day can give most of your plants a leg up against pests and disease.

Pest Management

Dealing with pests and diseases is a natural part of gardening. Even expert gardeners and farmers experience crop failure from time to time. **Prevention** is the first line of defense against pests, and is far easier (and more fun!) than dealing with pest outbreaks. As such, I don't use pesticide sprays in the garden, even when they are homemade. That's because many natural solutions can be at least mildly toxic to the soil life, if not just as toxic as chemical products.

Flowering cilantro attracts a variety of beneficial insects. It is planted here with strawberries.

Whether purchased or homemade, pesticides of any kind can kill beneficial insects (killing insects is their purpose, after all), alter the pH balance of the soil, leave a toxic residue on the crop, or destroy beneficial soil microbes. Soap-and-water spray, for example, is commonly used for natural pest control. But it might also kill beneficial soil microbes and change the soil pH, depending on the brand and dilution. Because I don't want to damage my garden ecosystem long-term or poison the food I plan to eat, I don't fight pests—I prevent them. If I fail at preventing them, then I learn from them, but I don't spray.

Prevent Pests in the Garden

The following techniques will help you prevent pest outbreaks naturally. The key to natural pest management is patience. For example, in the first year that we dug up our front yard and replaced it with an edible landscape, we had quite a few pest problems. I was really disappointed—we had put so much time and physical effort into creating the edible landscape. I wanted to save it from being devoured by pests! Instead of making a rash action, however, I waited, and continued to practice all of the following techniques.

> ONE OF THE WORST MISTAKES YOU CAN MAKE AS A GARDENER IS TO THINK YOU'RE IN CHARGE.
>
> *JANET GILLESPIE*

While we were doing our part, the beneficial soil microbes were busy getting acquainted to this new environment we had just created. That first year was a little like the Wild West, as the soil organisms duked it out and eventually came into a balance (or truce). We have seen progressively more improvement each year as the soil ecosystem has become more established.

The beneficial soil microbes help feed plants, keeping them healthy and well-protected against pests. If we had sprayed anything—even an "organic" pest solution—it would have disrupted their natural establishment period, delaying the balance we desired. It could have become a never-ending dependence on pesticides, when patience—and a willingness to feed the good guys what they require—is all it took. Here are some ways I have found to help prevent pest outbreaks.

Broccoli seedlings are planted with worm castings to improve plant and soil health.

Creeping thyme in garden paths repels pests and attracts beneficial insects.

A variety of flowers will provide habitat, pollen, and nectar for pollinators and beneficial insects.

Encourage healthy soil.

Healthy soil makes healthy plants with strong immune systems, which are better able to fight off diseases and pests. Organic matter, natural fertilizers, mulches, and no-till gardening will feed and shelter beneficial soil life.

Fertilizers are where I like to put my energy for building healthy soil. Worm castings are the richest known fertilizer and contain humus, the building block to soil. I also like to fertilize using plants called nutrient accumulators such as comfrey. These plants reach deep into the subsoil with their roots and gather nutrients. Used as mulch, their leaves fertilize the soil as they decompose. Fish and seaweed fertilizer is the only store-bought fertilizer I use. Diluted, I use it once a month in the garden to activate soil microbes. To read more about fertilizers and soil amendments that encourage a healthy soil ecology, see chapter 3.

Choose resistant varieties.

Give your garden a leg up by choosing plant varieties that are naturally resistant to pests.

Plant in the right place.

Plants will tolerate less than ideal conditions as long as they can, but over time they can become weaker and more likely to succumb to pests. Instead, reserve plants that need full sun for the full sun areas of your garden, and plant the leafy greens and root crops in partial sun areas. Likewise, be sure to match the water needs of certain plants to appropriate areas of the garden.

Give crops the proper watering.

When plants get too much or too little water, they will be stressed and more susceptible to catching a "bug." Avoid chlorinated water when possible by capturing rain in rain barrels. Chlorine can destroy beneficial soil microbes.

Attract beneficial insects.

Beneficial insects search for nectar, pollen, and shelter. Attract them to your garden by growing these flowers among the vegetables: calendula, coriander, parsley, and sweet alyssum. The following tall flowers and perennials will benefit the garden as an edge border: comfrey, coneflowers, cosmos, daisies, dill, sunflowers, and yarrow. Many beneficial insects prefer white and yellow flowers and will lay their eggs nearby, growing an army to patrol your garden for pests.

Repel pests.

Strong-scented herbs deter pests. Calendula, coriander, and garlic planted among the vegetables will help to deter pests, while anise hyssop, chives, lemon balm, sage, and thyme will benefit the garden as an edge border.

Rotate crops.

A good crop rotation practice will confuse pests and reduce their concentration in specific areas (see chapter 4).

Practice interplanting.

Interplanting means grouping together specific crops, herbs, and flowers to confuse pests as well as taking advantage of plant growth patterns to create more productive gardens. Pests enjoy monocrops, which is why industrial farms are often heavily sprayed with pesticides. Instead of monocrops, alternate rows of vegetables with rows of beneficial insect-attracting and pest-repelling herbs and flowers. For example, I like to interplant my cabbage family crops with cilantro and calendula to attract beneficial insects and repel pests.

Use floating row covers.

Summer-weight row cover fabric allows water and light to penetrate while keeping pests out. You may only need to use it over young plants until they're established. Weigh down the sides with heavy objects like bricks or rocks.

Or, consider adding permanent low tunnel hoops to raised garden beds for beans, beets, cabbage-family plants, chard, cucumbers, eggplant, melons, potatoes, pumpkins, spinach, and squash, which benefit from row cover to keep out Mexican bean beetles, cabbageworms, leafhoppers, leafminers, squash bugs, cucumber beetles, and flea beetles. Just be sure to lift the cover for a few hours each morning to allow bees to pollinate the plants.

Create permanent walkways.

Pathways of white clover, wood chips, or gravel each encourage different beneficial insects, while temporary pathways that are tilled each year destroy these insects and their habitat.

Common Pests and Beneficial Insect Controls

No solution is foolproof. Having a few pests is actually a good thing, because without them, the beneficial insects that dine on them wouldn't stick around! In the case of a pest outbreak, it will be essential to properly identify pests and beneficial insects—including the larval stages of each—before destroying any insect. The larval stages of each insect sometimes look completely different from the adult insect, and often it is the larvae that are the beneficial predators or pests, while the adults simply sip nectar from flowers.

Accidentally destroying beneficial insects reduces the ability of your garden ecosystem to self-regulate. Beneficial insects are attracted to gardens that have their favorite foods, so a few pests will draw them to your garden for a delicious treat. In other words, the occasional pest "bait" is okay. If it appears that a few pests have turned into a major outbreak, however, promptly remove diseased and infested plants and dispose of them in the garbage (not the compost pile) to keep the damage from spreading to other parts of the garden.

TIP: BENEFICIAL INSECTS

Go to www.TenthAcreFarm.com/tsmf-companion for a link to a database that identifies both beneficial insects and pests in different stages of their life cycles.

I look at a pest outbreak as an opportunity to learn about how I can strengthen my micro-farm ecosystem. Is there a mineral my soil is lacking, making my plants sick enough to catch a "bug"? If so, what organic material might supply it?

The "Common Garden Pests and Beneficial Insects" table gives a list of common pests, plants they attack, the beneficial insects that can help control them, and plants that will attract the beneficial insects.

To manage larger garden pests—such as deer and rabbits—it will take a variety of solutions.

First, know exactly who the culprit is. If you haven't seen an actual animal, you may think it's a deer when actually it's a groundhog. Sadly, this is a true story that happened to us at our community garden. We worked diligently—spending lots of money and time on materials—to install deer fencing and make it ever taller to deter our nemesis. We never actually saw a deer, and even though there were some convincing footprints, it wasn't until four years later when we saw the groundhog that we knew exactly what we were dealing with.

The praying mantis will hunt all manner of garden pests (and sometimes, other beneficials, too.)

Toads eat a variety of ground-based pests such as slugs and snails. They are attracted to moist, shady plant cover, and will need a water source to reproduce. Bordering your garden with flowerbeds and brush piles will provide the habitat they need.

Dragonflies will hunt primarily for flying garden pests. Keeping a birdbath with shallow water in it will help attract dragonflies and other winged allies.

Ladybug on asparagus—keep these beneficial ladies (and gents) around by planting some of their favorite foods like calendula, chives, cilantro, dill, and yarrow.

Yellow garden spiders look scary but are non-poisonous and beneficial. They are excellent hunters of all manner of garden pests, and are most active in late summer when the pests are most active, too.

COMMON GARDEN PESTS AND BENEFICIAL INSECTS			
Common Pest	Plants It Attacks	Beneficial Insects that Control It	Plants that Attract Beneficial Insects
Aphid	Most plants and trees	Green lacewing larvae hoverfly larvae ladybugs and their larvae minute pirate bugs praying mantis	Coreopsis white cosmos dill yarrow
Cabbage looper	beets, cabbage family, celery, lettuce, peas, spinach, tomatoes	tachinid flies	cilantro dill parsley
Cabbageworm	cabbage family	Green lacewings spined soldier bugs tachinid flies	cilantro goldenrod sweet alyssym
Corn earworm/ Tomato fruitworm	beans, cabbage, corn, okra, peppers, squash, tomatoes	Assassin bugs, lacewings, minute pirate bugs, parasitic wasps	Queen Anne's lace, white cosmos, yarrow
Leafhopper	beans, eggplant, fruit trees, potato, squash	Assassin bugs, ladybugs, minute pirate bugs	Fennel, goldenrod, yarrow
Squash bug	cucumbers, melons, pumpkins, squash	Tachinid flies	Cilantro, dill, parsley

A game camera can be priceless for this. If you've discovered an animal, but aren't sure what it is (Mole? Vole? Shrew? Mouse?), do some research online to identify it. They could be after completely different things in the garden. For example, moles are carnivorous and are likely chasing tasty insects that live in your lively garden, while voles are the vegetarians who are chewing on roots and destroying crops.

A farm cat can help ward off these smaller invaders. Also, raised beds can deter pests like rabbits who prefer to be low to the ground for camouflage and easy escape. Grow crops they don't like. For suggestions on what to plant (or not) around deer and rabbits, see chapters 4, 10, and 11. But ultimately, it may come down to species-specific fencing and repellents. As we discovered with our deer-groundhog confusion, the proper identification of species is priceless.

TIP: TRACKING PESTS

Keep notes of what pests you encountered, what plants they were attracted to, when they showed up, what methods of prevention or treatment you tried, and whether or not those actions kept them at bay. There is room to take notes about pest problems on each monthly checklist. Download the checklists at www.TenthAcreFarm.com/tsmf-companion.

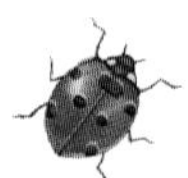 Guide to Harvesting Vegetables

The moment you've been waiting for! Your vegetables are taking on the likeness of those that you're accustomed to seeing at the farmers' market or grocery store, but are they ready for harvest? This harvest guide will help you pick your vegetables at just the right time. This guide includes general harvest windows, which may differ slightly in your region. Season extension techniques—such as cold frames—will also affect your harvest times. See chapter 4 for tips on planting, growing, and extending the growing season.

TIP: KEEP A HARVEST LOG

Before beginning to harvest, download my Harvest Log at www.TenthAcreFarm.com/ tsmf-companion. I've kept harvest records for the last six years, and I'm so glad I did. I've been able to see what my best and least performing fruits and vegetables are, and track accurate harvest windows for each crop based on the specific conditions of my yard. Knowing my true harvest windows has helped me plan vacations around certain harvests so we don't miss out.

Asparagus

Asparagus: Harvested February through July, depending on your location. Harvest lightly in spring, two years after planting. Leave half of the spears to grow out to ferns. In the third year, harvest throughout the entire three-month harvest window. Harvest when the spears are about seven inches in length by bending the spears at the base until they snap. Or cut with a knife just below the soil.

Beans: Harvested starting around 50 days after planting. Harvest as soon as the pods have filled out. Waiting too long will result in tough pods. Harvest daily to keep the plants producing; otherwise production will slow.

Beets: Harvested starting around 40 days after planting. Beets have the best flavor if harvested after a frost at golf ball size. For storing, harvest at tennis ball size. Store beet greens and beets separately for best results.

Beets

Broccoli: Harvested starting around 50 days after planting seedlings outdoors. Harvest before the flower buds open, when the buds are still tight. Cut the stalk at a 45-degree angle about five to six inches below the base of the head. After the central stalk has been cut, continue to harvest side shoots regularly.

Brussels sprouts: Harvested starting around 90 days after planting seedlings outdoors, tasting better after frost. Begin picking at the bottom; upper sprouts will continue to mature. Around mid-September, cut the top of the plant off by four inches. This removes

TIP: VEGETABLE STORAGE

Think ahead about the type of storage you will need for the vegetables you harvest. Some will store fine in the refrigerator or on the kitchen counter, while others will need cool, dry storage like that found in a root cellar or basement.

the buds that wouldn't have time to mature before harvest, resulting in larger existing sprouts. Harvest individual sprouts when they are one inch in diameter by snapping them off.

Carrots

Cabbage: Harvested starting around 65 days after planting seedlings outdoors. Young heads store the best. Harvest anytime the heads are solid and firm.

Carrots: Harvested about 50 days after planting. Harvest at any time after the carrots have developed a dark orange color and are at least ½-inch in diameter (they are a pale color when young). Carrots are sweeter after a frost. Use a trowel or other digging tool to harvest, pulling may snap the carrots in half. You can store carrots in the ground up until the ground freezes. Cut the tops off before storing.

Cauliflower: Harvested starting around 50 days after planting seedlings outdoors. Harvest when heads are firm and curds are solid, around six to eight inches. If the curds become loose, the ideal harvest window has passed.

"Ruby red" chard

Celery: Harvested starting around 80 days after planting seedlings outdoors. Harvest anytime the stalks are an edible size by cutting individual stalks or pulling the entire plant.

Chard: Harvested starting about 50 days after planting or seeding. Harvest outer leaves first, best at eight inches long or smaller. Plants will continue to produce after harvesting.

Collard greens

Collard greens: Harvested starting around 50 days after planting seedlings outdoors. Best after a frost; harvest individual leaves starting with the outer leaves. Plants will continue to produce after harvesting.

Corn: Harvested starting around 65 days after planting. Harvest when the ear silks have dried and turned brown, and kernels are milky when a kernel is punctured with a thumbnail. Pull the ear downward and snap off the ear, twisting if necessary. Check daily; the harvest window for great taste is short. Prepare corn within 24 hours of harvesting for best taste.

"Chesnok red" garlic

Cucumber: Harvested starting around 50 days after planting. Harvest cucumbers daily when they are six inches long, or three inches long if you plan to make dill pickles. You will enjoy better tasting cucumbers and a more productive plant when you harvest often.

Eggplant: Harvested starting about 58 days after planting. Harvest when eggplants are four to eight inches long. Younger, smaller eggplants are better tasting and harvesting them young keeps the plants producing. Use hand pruners to harvest rather than pulling.

Garlic: Harvested June through September, depending on your location. Harvest when tops begin to yellow, but still have three to four green leaves on the stem. Harvest when the soil isn't waterlogged. Lift the bulbs with a garden fork rather than pulling. Cure them in a dry, cool, shaded, well-ventilated spot for two weeks.

If you grew a hardneck variety of garlic, cut the stems off one inch above the bulbs and trim the roots off before hanging the bulbs in a mesh bag. Softneck garlic can be hung in bundles or braided and then hung in a cool, dark location.

"Red Russian" kale

Kale: Harvested starting about 50 days after planting seedlings outdoors and is best after a frost. Harvest individual leaves starting with the outer leaves. Plants will continue to produce after harvesting.

Kohlrabi: Harvested starting about 40 days from planting. Kohlrabi is sweet and tender when picked young (two inches) and after a frost, but pick anytime the kohlrabi is between two and five inches in diameter.

Leeks

Leeks: Harvested starting about 75 days after planting seedlings outdoors. Harvest anytime the stem is at least an inch in diameter or larger. Store and use only the white section.

Lettuce: Harvested about 40 days after planting seedlings outdoors. Leaf lettuce varieties can be harvested at any time. Pick the outer leaves as needed, or cut the entire plant an inch above ground 30 days after direct sowing in the garden. If leaf lettuce is cut, leaving the roots in place, it will grow back. This is called cut-and-come-again.

"Parris Island" Romaine lettuce

Melons: (cantaloupe, watermelon) Harvested about 80 to 100 days after planting by seed. Knowing when to harvest melons isn't clear. It's important to look at the seed packet of the variety you're planting for the "days to maturity," which is the days from planting to harvest. Mark it on your calendar!

Cantaloupes soften slightly at the blossom end when ripe, while watermelon rind loses its gloss, the nearest tendril turns brown, and the side touching the ground changes from white to creamy yellow. Most melons must be cut from the vine with hand pruners.

Okra: Harvested starting around 50 days after planting seedlings outdoors. Harvest when pods are two to four inches long to avoid tough pods and to keep the plant producing. Snip off the pods with hand pruners.

Onions

Onions: Harvested about 100 days after planting seedlings outdoors. Harvest scallions when the green onions are about 6 inches tall and the thickness of a pencil. To harvest bulb onions, wait until the onions have developed skin and the tops are falling over. Harvest on a day when the soil isn't waterlogged.

Sugar snap peas

Bell peppers

"Cherry belle" radishes

Cure onions in the sun for two days, then move them to a dry, partially shady area to continue curing for two to three weeks, until the necks have thoroughly dried out. Cut the tops off about an inch above the bulbs, trim off the roots, and store them in a well-ventilated, dry, cool, dark location.

Peas: Harvested starting around 50 days after planting. Pick peas just when the pods are full, except snow peas, which are better when they are younger and thinner. Pick often and process the peas within a few hours of harvesting.

Peppers: Harvested starting around 50 days after planting seedlings outdoors for green peppers and 70 days after planting for red peppers. Pick peppers any time after they've reached full size. Green bell peppers are ready to pick when they're around four inches wide. Picking them early will encourage the plant to produce more peppers. If left on the plant, green peppers will mature to red peppers. After the first picking of green peppers, continue to pick green peppers or wait until they turn red.

Potatoes: Harvested May through July for spring planted and August through December for summer planted, depending on your location. When they're ready to harvest, the foliage will die down. Leave potatoes in the ground for two weeks after this time to toughen the skins for storage. Do not harvest when the soil is waterlogged. Dig tubers up with a spade fork and dry them in the sun for a day. Brush off the soil, then cure them in a cool, dry, well-ventilated, humid area for two weeks. Store them at 45 degrees F.

Radishes: Harvested as early as 21 days after planting. Harvest them for fresh eating when roots reach the diameter of a quarter for a milder taste. As they mature, they'll develop the spicy radish flavor. Store green tops separately from roots.

Rhubarb: Harvested April through June, depending on your location. Only the stems of this perennial plant are eaten; the leaves are poisonous. In the second year, harvest half of the stems that have reached 10 inches long by pulling and twisting at the same time.

Summer squash

Rutabaga: Harvested starting about 90 days after planting. Wait until there have been at least two good frosts. Roots three to six inches are ready for harvest. The smaller roots will be sweeter. Cut the tops off before storing.

Spinach: Harvested starting about 35 days after planting. The largest and sweetest leaves are harvested after a frost. Cut spinach leaves at any size; the plant will continue to produce after harvesting. Or cut just below the root attachment to harvest the whole bunch. Spinach planted in August will produce plants that can overwinter.

Sweet potatoes

Summer Squash: (yellow squash, zucchini) Harvested 48 days after planting. Best tasting when smaller. Harvest when fruits are just four to eight inches long, and harvest often. This will also keep the plant productive.

Sweet Potatoes: Harvested around 90 days after planting. Harvest sweet potatoes before a hard frost, and before soil temperatures dip into the 50s. Use a garden fork to gently dig the tubers. Cure for 10 days in a dark room at 80 degrees F with 90% humidity. I've had success curing them in a plastic bag on a sunny window, covered with a towel. Curing toughens the skin for storage and sweetens the flavor. After curing, store at 60 degrees F in a dry spot.

Tomatoes

Tomatoes: Harvested starting about 60 days after planting seedlings outdoors. Pick when fully colored for best flavor. Pick green

Turnips

if frost threatens. They're edible green, or let them ripen indoors above 60 degrees F, not in the refrigerator.

Turnips: Harvested 30 to 50 days after planting. Harvest before frost and store roots separate from the greens.

Winter Squash: (acorn, butternut, pumpkin) Harvested starting around 85 days after planting. Winter squash is ready for harvest when the skin hardens and cannot be punctured by the thumbnail. The stem will dry. Cut the stem two inches from the fruit after a light frost and before a heavy frost on a dry day. Cure fruits in a warm place (80 degrees F) for two weeks. This hardens the skin for storage. Once cured, winter squash will keep up to four months in a dry location at 55 degrees F.

Cleaning up the Garden for Fall

When fall comes around, it signals to most of us that the season is winding down. It's at this point that I'm ready for a break. It would be easy to walk away until spring, leaving the garden just as it is. In some cases, this isn't a terrible idea. As long as there weren't pests or disease, leaving spent plants in the garden can support overwintering beneficial insects, as well as protect and build soil.

However, there are a number of things that you can do to put the garden to bed for winter that will save time in the spring. No one wants to start the garden season feeling behind!

Record your garden layout and notes.

If you haven't recorded your garden layout yet—what you planted where—now is the time to do it before spent crops are removed. Also, make any notes about this year's experience. Perhaps the kale didn't grow well and the tomatoes were diseased, but the beets were abundant. Taking the time to make notes now will make it easier to plan next year's crop rotation for reducing pests and avoiding nutrient deficiencies.

Make a fall harvest.

You've worked hard in the garden all season. Don't let those last harvests go to waste! Take advantage of the earth's bounty before cleaning up.

Save seeds.

Harvesting seeds is a different kind of harvest, but just as useful as harvesting the produce. Saving your own seeds from crops and flowers that produced well will help you develop varieties that are adapted to your specific growing conditions.

Remove spent plants.

There are two schools of thought about spent plants: One is that they should all be ripped up and added to the compost pile to create a clean slate and reduce the workload for next year's spring garden. The other school of thought is that spent plants should be left in place to biodegrade. I take a middle-of-the-road approach: I cut plants off at the base, leaving the roots intact and adding the spent plants to the compost pile. This way, soil amendments and mulch can be easily added to the beds, but the roots are left in place to enrich the soil.

The roots are like bait. Instead of luring fish, though, the root baits worms and other soil organisms that enjoy eating the roots and helping to decompose them in the soil. By the time spring rolls around, the area around the old root tends to be higher in nutrients and—because of the worm poop—better at regulating moisture, too. The decomposing root will feed the new plants or seeds that are planted near it and help them to be more resilient.

It took me some time to figure out how to leave roots intact and still plant in square-foot squares or garden rows. I discovered that it works fine to plant my seedlings around the old roots, even if they are a little off-center. Those seedlings always seem to do better than the seedlings not planted next to an old root! For direct sowing, there may be a little circle of unplanted space around the old root, but the veggies seeded right next to it will get bigger than those not directly next to it.

There are two additional considerations when cleaning up plants. Plants that were diseased or bothered by pests should be pulled up completely and discarded in the garbage or burned in order to reduce overwintering of "bad" organisms. Also, many gardeners leave the roots of legumes (beans, peas) intact in order to enrich the soil with nitrogen. If you would like to experiment with this, simply cut those plants at the base, leaving the roots and adding the plant matter to the compost pile.

Weed the garden.

Now that the garden is cleared out, it's easy to spot and remove weeds, another way to save time in the spring. Most weeds will make an excellent addition to the compost pile, but there are a

few types of weeds that I cut off at the base and leave right on top of the soil. Weeds like clover, dandelion, chickweed, lamb's quarters, plantain, and purslane will actually enrich the garden soil with nutrients as they decompose.

Loosen the soil with a digging fork.

This is a helpful step for maintaining the no-till garden. Loosening the soil (without turning it over) will aerate it and improve its absorption capacity of winter rain. This isn't a good time to till, however, as that practice can increase erosion by destroying worm tunnels and fungal networks that help to hold the soil together over the winter.

Add finished compost and maintain the compost pile.

It's a good idea to do some maintenance on the compost pile before it freezes over the winter. This works out well, because it is also a good time to add finished compost to the garden to prepare for spring planting. Add any additional soil amendments to the garden along with the compost. Now turn the compost pile with the added weeds and spent plants. This aeration will help to speed up the composting process.

Sow a cover crop, lay dried herbs, or apply manure.

Winter cover crops fertilize the garden, increase microorganism activity, reduce erosion, break up compacted soil and aerate it, and outcompete early spring weeds. Many types of cover crops will die back on their own, making spring planting easy, while other cover crops will need to be cut back and turned into the soil about three weeks before planting. The organic matter will help to condition the soil. No-till gardeners should seek out non-grass type cover crops that are more easily turned into the soil by hand.

Fresh or dried herbs and livestock manure can also be applied in the fall and will fertilize, condition the soil, and increase biological activity.

Mulch garden beds with shredded leaf mold or straw.

Mulching garden beds will help to protect soil microorganisms, worms, and overwintering beneficial insects. It will also help to prevent erosion. Leaf mold—shredded leaves that have composted for at least a year—is ideal, but straw can be used if leaf mold isn't available. Seek out

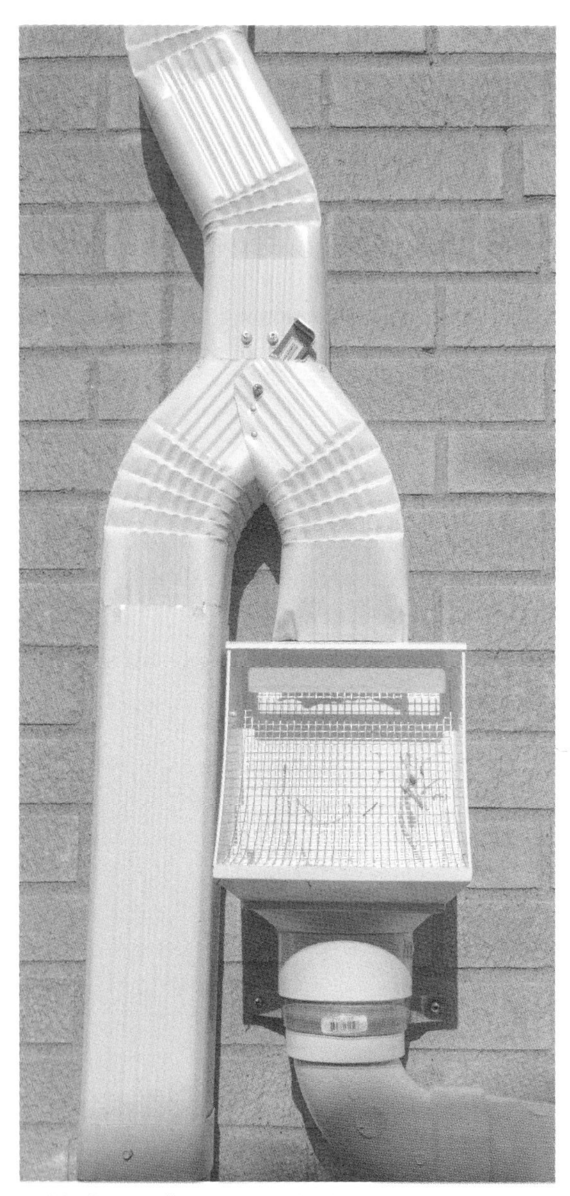

This diverter allows water to be directed toward the rain barrels or toward the sewer, which can be helpful in times of heavy rain and in freezing winters.

Empty rain barrels for the winter in areas with freezing winters to keep them from cracking. Divert the water back to the storm sewer, or if you have a rain garden, keep the valves open to keep barrels drained.

Finished compost

chemical-free straw from organic operations, since conventional straw is known to contain herbicide residues that can be persistent, reducing germination rates and plant vigor.

Cover crops will not need to be mulched, but be sure to mulch vacant annual beds as well as overwintering garlic and other vegetables. Perennials should also be mulched with leaf mold, straw, or wood chips.

Now is a good time to rake up this year's leaves, shred them, and begin to make leaf mold for the following year.

Create new garden beds.

Building garden beds in the fall is the best thing you can do for starting the spring garden without a hitch. Garden beds that can settle and mature over the winter will be more productive and resilient in their first season. Use a digging fork to aerate the soil, which will encourage worms and other soil organisms to till the following amendments into the existing soil. Lay cardboard over the proposed garden area. Collect soil amendments such as worm castings, finished compost, manure, coffee grounds, grass clippings, etc. and spread a nice thick layer over the area. Top it with three to six inches of compost soil. Seeding it with a cover crop would be one way to help establish the garden bed, or mulch it well.

Store, build, and repair.

Store garden ornaments, plant signs, trellises and supports, garden tools and equipment out of the elements. If it freezes where you live, empty your rain catchment containers onto perennials. Make structural repairs, build new infrastructure such as a shed or swale, or sheet mulch areas that were overtaken with weeds.

These tips will improve the health of your garden and give you a head start on next year's spring garden.

Easy Seeds to Save

Saving your own seeds can be easy and fun, and it is a great way to save money, increase your self-reliance, and begin crafting varieties that are well-adapted to the conditions of your garden. The easiest vegetable seeds to save are beans, peas, peppers, and tomatoes.

Shredded leaf mulch ready for the garden.

A fall harvest

To save seed from the vegetables you grow, make sure the variety you're growing is an open-pollinated variety rather than an F1 hybrid. F1 hybrids are commercially produced varieties specifically bred for pest or disease resistance, or increased yields. Hybrid seeds either produce a child plant that is not true to the parent, or they produce sterile seeds that will not reproduce at all.

Another important consideration for saving many types of vegetable seeds is separating same-crop varieties in the garden to reduce the possibility of cross-pollination. Cross-pollination occurs when pollen is exchanged between flowers of different plants by pollinators. It sometimes results in the seeds producing a crop that is a cross. For example, some plants, like beans, need only have different varieties separated by 150 feet, while corn varieties on the other hand, need to be separated by 1,000 feet. I've planted sweet peppers and hot peppers beside one another only to discover that they cross-pollinated, resulting in the sweet peppers *looking* like sweet peppers, but *tasting* like hot peppers. Now that's shocking!

Bean seeds saved

Bean and Pea Seeds

Separate bean varieties by 150 feet and pea varieties by 50 feet in the garden. For both, allow the pods to dry on the vine and turn brown. Harvest the dry pods after a period of dry weather, and lay them on a wire mesh screen or wire shelf to dry indoors for two more weeks in a well-ventilated area. Shell the pods. Reserve the seeds and compost the pods. Store and date/label.

Bean and pea seeds are especially susceptible to bean weevils, which lay eggs that hatch in the seeds while in storage and destroy a seed collection. To avoid damage, simply freeze the seeds for five days. See "Freeze for Long-term Storage" in this chapter for instructions on how to safely recover seeds from the freezer.

Pepper Seeds

Separate varieties by at least 50 feet and up to 400 feet if you have the space in the garden. Harvest an unblemished, fully mature pepper. I know it's hard not to eat the prettiest ones! Spread the seeds on a paper coffee filter to dry until seeds break when folded. Store and date/label.

Tomato Seeds

Separate varieties by at least 10 feet and up to 100 feet in the garden. Harvest an unblemished, fully ripe tomato. Cut open the tomato and scoop out the seeds and pulp. Place the seeds and pulp in a glass with a small amount of water in a warm spot (60 to 75 degrees F) for about three days. *This fermentation process will not smell good!* Warmer temperatures will speed up the process. The process prepares the seeds for storage by destroying diseases and increasing the seeds' ability to germinate when it's time. Stir every day.

After a layer of mold covers the surface, add more water to double the amount in the jar and stir well. Once the contents settle, pour off the water, pulp, and immature seeds that are floating. Retain the viable seeds that have settled at the bottom. Repeat the process of adding water, stirring, and pouring off debris until the water is clear. Now strain the seeds and lay them on a paper coffee filter in a single layer to dry in a warm, dark location. Stir the seeds once a day until they are fully dry, about two or three days. Store and date/label.

Herb and Flower Seeds

In order to save seed from herb and flower species, there's one easy requirement: They must first be allowed to flower. As the flower matures and dries in the garden, seeds in the flower head get ready to break away from the plant and scatter. Collect dry seeds during a dry spell, and spread them on a framed window screen to continue to dry indoors for two weeks. Store and date/label.

Cilantro seeds, also called coriander

Storing Seeds

If you've saved your own seeds, be sure that the seeds are thoroughly dried before storing. Here are some tips on properly storing seeds and estimating seed viability.

Dry Storage

Airtight containers are important for storing seeds—the containers can be glass, metal, or plastic. I store my seeds in seed envelopes in a large, airtight, Tupperware-type container. I also like to use Mason jars, but I trust myself less with them when I go outside to plant. I'm always worried that I'll drop and shatter glass in the garden! I also save the silica gel packets that come in shoe boxes and vitamin bottles and add them to my seed-saving containers as a safeguard against moisture.

Seeds should be stored in a dry, dark place with consistently cool temperatures—like a cupboard. I store my seeds in the dark basement.

Freeze for Long-term Storage

For long-term storage—or if you don't have a basement or cupboard with consistent temperatures—consider freezing (completely dry) seeds in a glass jar. The refrigerator is second-best, since temperatures aren't as consistent there.

> THE TWO GREATEST ENEMIES OF STORED SEEDS ARE HIGH TEMPERATURE AND HIGH MOISTURE.
>
> *SUZANNE ASHWORTH, SEED TO SEED*

To recover seeds from the freezer for use, set the jar out at room temperature for 12 hours for it to reach room temperature. This will prevent moisture from condensing on the seeds (moisture = enemy #1)! Then expose the seeds to air for a few days before planting. Refrain from moving seeds from the freezer to room temperature more than once, as each transfer will reduce the viability of the seeds.

Organizing Seeds

Having a system for keeping seeds organized allows you to take stock of what seeds you have, so you don't end up buying seeds you already own. It can also help you keep track of your seeds by age, so that older seeds are used first, and expired seeds are composted instead of planted.

Each seed packet has an origination date on it. This is the date the seeds were collected. The first thing I do is use a permanent marker to write this date on each seed packet. Even though the date is already on there, I write it really big, so as I'm sifting through seed packets I can quickly categorize them by year if I need to.

There are two ways I like to organize seeds: card catalog style and Mason jar style. Your storage organization will largely depend on your lifestyle and environment, but no matter which style you choose, be sure it is an airtight solution that keeps seeds dry and cool. Don't forget your silica gel packets!

A plastic shoe box transforms into a card catalog style seed organization box.

Card Catalog Style

Are you old enough to remember library card catalogs? Use a rectangular airtight container that is deep enough to store seed packets standing up. Make sure the lid fits well when full. I love this method because all the seed packets live together in one or two containers. Dividers can help to find things even quicker. Seeds of a certain type can be catalogued in order of their origination date, so older seeds get used first.

All of my lettuce seed varieties fit into this gallon size mason jar.

Mason Jar Style

Mason jars allow you to store seeds in smaller units. For example, you could store short-lived seeds in the freezer so you can ensure their viability until the next

growing season, while longer-lived seeds may store okay under the bed or in a cupboard. How you organize your Mason jars will depend on your storage needs and how many seeds you're storing. If you're a tomato lover, store all of your tomato varieties together in one Mason jar. Or, try themed gardens in Mason jars (which make great gifts!). Seeds for a spring garden, a salad garden, or a salsa garden could store well together in one jar.

However you organize your seeds, be sure to keep track of seed origination dates.

SEED VIABILITY		
Short-lived Seeds (1 to 2 years)	Intermediate Seeds (3 to 4 years)	Long-lived Seeds (5 to 6 years)
okra	bean	cucumber
onion	beet	lettuce
parsley	cabbage family (broccoli, Brussels sprouts, cabbage, cauliflower, kohlrabi, etc.)	radish
parsnip	carrot	
pepper	celery	
sweet corn	eggplant	
	leek	
	pea	
	pumpkin	
	spinach	
	squash	
	tomato	
	turnip	
	watermelon	

TIP:

Toss expired seeds into a wild section of the garden. Sometimes, they produce a surprise crop! At our community garden, we had a "Seed Toss" each year at the season kick-off party. It was fun to see what would grow in the wild area!

Seed Viability

As seeds age, their germination rate naturally declines. All seeds will stay viable for at least a year, and with good storage practices, you can expect many of your seeds to remain viable even longer.

There are certainly quite a few things to do in between planting and harvesting to keep in mind, such as checking on the garden regularly, mulching, watering, keeping an eye out for pests, and taking good notes. This labor of love is what seals the deal for a successful harvest. Cleaning up the garden for fall and saving seeds properly are two end-of-season topics that don't get talked about nearly enough, but they can really set you up for an even better garden next year.

I WOULD RATHER BE
ON MY FARM, THAN BE
EMPEROR OF THE WORLD.

GEORGE WASHINGTON

PART 3: ADVANCED MICRO-FARMING TECHNIQUES

10 EDIBLE LANDSCAPING

THE BEST TIME TO PLANT A TREE WAS 20 YEARS AGO.
THE SECOND BEST TIME TO PLANT A TREE IS NOW.

CHINESE PROVERB

Suburban micro-farms will come in all shapes and sizes, in all types of neighborhoods. While some of us prefer function to aesthetics, other micro-farmers will seek beauty for their own enjoyment, or as a way to comply with aesthetic expectations of their neighborhood. Either way, growing perennials such as fruit and herbs is an excellent strategy for increasing yields without more work, as well as maintaining aesthetics in the landscape. The following are some of my ideas for creating an edible landscape and utilizing fruit and herbs on the micro-farm. For even more ideas on using edible perennials on the micro-farm, see chapters 5 and 11.

Edible landscaping is an easy way to grow food while keeping a more traditional appearance in the front yard. It is also a great start for both beginners and busy micro-farmers. I began my edible front yard by simply tucking a few vegetable plants into the regular landscaping.

After much trial and error in the aesthetic department, my edible front yard now contains a currant bush hedge lining the front porch, black raspberries bordering the front of the house, strawberries on a berm that captures water from the roof, cherry trees lining the sidewalk in the parking strip, and a variety of edible herbs and flowers.

I wasn't always settled on this arrangement, however. Our edible landscaping went through an awkward phase as I experimented with different ideas. For example, the strawberry berm that runs through the middle of the yard capturing water from the roof was originally planted with asparagus. I hadn't thought about how ridiculous it would look to have a tall line of asparagus running through the middle of the front yard! But as soon as it grew tall, it was clear I had made a mistake.

Throughout this chapter, I will share some of the things I've learned over the years of keeping an edible front yard.

 Why Landscape with Edibles?

Simply put: We can't eat lawn. When I found out that lawn is the largest crop in the United States—and realized how abundant it is in the suburbs—I decided to give my yard a makeover! Suburbanites typically spend their weekends maintaining an ornamental lawn and pointless landscape plants, while edibles could be more beautiful, produce an edible yield, and—with the right crop choices—require not much more maintenance.

While I have a particular fondness for a productive landscape, I also enjoy a beautiful one that attracts beneficial insects and is friendly to wildlife. Edible landscaping is a softer, gentler approach to growing food.

My edible front yard (Photo by Ken Stigler Photography)

While the development and maintenance of a lawn is relatively mind numbing (not to mention potentially polluting), what I have found after years of micro-farming my front yard is that I'm more alive and engaged with the edible landscape. For example, when the strawberries, cherries, or black raspberries ripen, it is an exciting moment! Nothing about a lawn is that exciting.

An edible landscape gave me a reason to interact with my neighbors. While the usual dog walker or passerby will exchange niceties about the weather, the edible landscape gave us something interesting to talk about. *Edible Estates* is a favorite book of mine that documents the transformation of lawns across the United States into productive landscapes, and also explores the social aspects of increasing interactions with neighbors.

Edible landscaping is often talked about as a modern marvel thanks to pioneers like Rosalind Creasy, author of *Edible Landscaping*. But landscaped gardens containing edibles have been the norm throughout history from English cottage gardens all the way back to ancient Persia. Edibles truly make supremely beautiful landscapes.

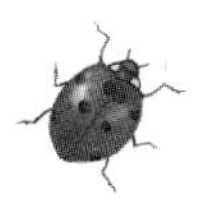 Things to Consider before Breaking Ground

Just as with other micro-farming techniques, you will need to choose solutions that work for your unique growing conditions, time availability, and budget. Here are some additional things to consider before finalizing your design.

Deer, Kids, and Dogs

What do they have in common? They all take the most direct path from point A to point B. If garden beds are easy to step in or walk through, then they probably will be by this bunch! Create diversions and obstacles that direct traffic away from your prized plants. When we planted edibles in the parking strip (between the sidewalk and street), we planned accordingly so it could handle the steady stream of traffic by dog walkers, kids on bikes, and parked-car passengers.

Time

How much time do you have to commit to maintenance? Ripping out lawn and replacing it with something else will—sadly—require a bit more maintenance to keep tidy. If time is not on your side, then consider leaving the lawn intact and simply replacing the existing landscaping plants with edibles.

Take it from me: Annual plants will need more attention than perennials to retain an aesthetic appeal. As I began our edible landscape journey, I had no idea how much time it would take to keep the annual vegetables looking tidy and weed free—an important consideration for the front yard landscape. You'll need a plan to fill the space after vegetables are harvested, or you will be left with gaping holes in the landscape. This is why I have transitioned our edible front yard to contain mostly perennials.

Money

A long-term landscape will be made up of mostly perennials, but plants are expensive. If you don't mind bare ground for a while, buying seeds and young starts will be much cheaper. We built our landscape over several years to spread out the cost of buying plants. We also used fast-growing annuals to fill in the spaces until perennial plants reached their mature size.

The curved edge of the strawberry berm—lined with stone—defines this front yard garden.

A block planting of yellow California poppies forms a backdrop to the front yard strawberries.

Landscape Design Principles for Visual Appeal

There are a few design principles that professionals use when designing new landscapes. If you take them into consideration, you can landscape like a pro!

Line

A line defines a space and connects people to the landscape. The line could be a curved walkway or an edge of the garden, anywhere our eyes are drawn to follow a line or edge.

Unity

A unified grouping of plants creates order and attracts attention from both humans and beneficial insects. Group plants of the same type together rather than alternating colors or textures. We planted the flower garden with groups of bold colors as a backdrop to the edibles.

Emphasis

A specimen tree creates a focal point. Instead of a Japanese maple, dogwood, or other common specimen tree, try a fruit tree instead.

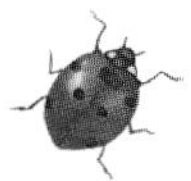

Planning Techniques for the Edible Landscape

Some popular ways to add edibles to the landscape are planting an edible hedge or privacy screen, or simply matching the colors of various vegetables, herbs, and flowers in a visually appealing way. Let's take a look at each of these techniques.

Plant An Edible Foundation Hedge

A foundation planting simply hugs the foundation of a building. As I've mentioned before, we planted currant bushes as a foundation hedge lining our front porch because they are shade tolerant, easy to grow, have beautiful berries (red currants), and are fragrant when brushed against (black currants). While the bushes were young, we grew a "hedge" of broccoli to fill the space!

A hedge of currant bushes lines the front porch.

A black raspberry hedge lining the front of the house adds color to the front landscape when blooming and fruiting.

We planted black raspberries underneath the front windows because they are shade tolerant, have a clumping habit (they won't "walk" around the yard), have beautiful red canes in winter, eye-catching fruit, and are relatively easy to prune. They are also thorny, making them a good security planting.

Other edible foundation hedge plants include bush cherry, gooseberry, and rugosa rose, which can all be pruned for tidiness. Did you know that rugosa rosehips have 50% more vitamin C than an orange?

Edible Privacy Screens

Living in suburban or urban environments, sometimes we need to create a little buffer for privacy. Edible plants can help create a living screen that will last much longer than a fence and create more biodiversity. As I've mentioned before, we used dwarf cherry trees in our parking strip, and we did so partially because their dense foliage created a soft buffer between our front yard and the street. They also have beautiful white flowers in the spring, gorgeous fruit that looks like Christmas tree ornaments, and interesting bark.

Other edibles for privacy screening include asparagus, elderberry, Nanking cherry, and serviceberry. If you have the space for them, chestnut, hickory, pawpaw and persimmon are tall trees with dense foliage for privacy.

Serviceberry bushes are beautiful when blooming.

Color Schemes in the Edible Landscape

I've enjoyed giving my edible front yard an annual color scheme. Here are some of the themes that I've used over the years. Mix and match, too!

COLORS IN EDIBLE LANDSCAPING				
Purple	Yellow	White	Red	Green
"Red Russian" kale	yellow chard	cilantro	cherry tomatoes	basil
"Rosa Bianca" eggplant	yellow bell peppers	garlic chives	cayenne peppers	broccoli
"Purple Beauty" bell pepper	yellow cherry tomatoes	oregano	red chard	kale
chives	yellow California poppies	sweet alyssum	red California poppies	collards
			nasturtium	parsley
				sweet potato
				zucchini

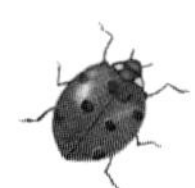 Choosing What to Plant

While there are many edibles to choose from when designing the edible landscape, this section will go deep into vegetables, fruit, herbs, flowers, and more, to help you choose exactly what will do best for your situation.

Vegetables in the Edible Landscape

I've learned that vegetables are tricky in the edible landscape. One year, I planted garlic in the front yard rain garden, and it was beautiful. I harvested it mid-June and immediately seeded the

area with winter squash (which I thought would make a beautiful ground cover). But multiple sowings never took. The bed—which was center of the front yard—remained empty all season.

Root vegetables are also tricky in the edible landscaping because they sometimes have trouble germinating. Suddenly it's not only a question of whether I'll get a harvest, but also what to do about the bare spot in the landscape.

Because of many examples like these, I avoid single-harvest crops—such as root vegetables—altogether in the edible landscape. Rather, I only plant vegetables in which the fruit or leaf is harvested, because the plant remains intact and continues to produce throughout the season. Examples are cherry tomatoes and peppers—both reliable in the edible landscape—as are cut-and-come-again leafy greens such as chard, kale, and collard greens.

My favorite vegetable and flower combinations include cut-and-come-again leafy greens: chard with sweet alyssum, kale with Johnny jump-ups, or collards with garlic chives.

The Chard and Sweet Alyssum Combination

Chard is a leafy green that produces large leaves with bright, colorful stalks. It remains visually stunning all season long because it is heat- and drought-tolerant, and attracts few pests. The cut-and-come-again nature of this plant allows the leaves to be harvested as desired, while the plant remains intact. I like to grow it along the walkway in my edible front yard. There are many varieties—we've grown the orange stem, ruby red, and bright lights varieties.

Red Russian kale (with its purple stems) goes nicely with purple coneflower in the edible landscape.

The low-growing habit of sweet alyssum is both an effective living mulch and a great border plant. The white-flowered variety works great with any variety of chard and will attract the most beneficial insects. To read more about this glorious edible landscaping combination, see chapter 9.

Edible Landscaping with Fruit

Lee Reich, author of *Landscaping with Fruit*, states, "The time has come for fruit trees, shrubs, and vines to come into their own as ornamentals." He calls this "luscious landscaping." Fruit trees and berry bushes produce beautiful

flowers in the spring. Some fruit trees to try in the landscape are American persimmon, apple, cherry, cornelian cherry, mulberry, pawpaw, peach, pear, and plum. Some fruit-producing shrubs for the landscape are currant (of course), blueberry, gooseberry, goumi, Nanking cherry, rose, and serviceberry.

Patience is a virtue here: It takes a couple of years for new perennials to establish themselves, fill in the space fully, and produce fruit.

Strawberries

Here's what I've learned about strawberries: pretty much everyone loves them. They are shareable and bring a smile to kids and neighbors alike. We grow strawberries in our front yard because they are extremely productive, they don't take up a lot of space, their deep roots can stabilize a slope and slow drainage, and they make a nice ground cover.

In the edible landscape, though, be wary: regular, June-bearing strawberries create runners that will "jump" out of the bed and "walk" away to plant themselves in pathways and other areas where you don't want them. Suddenly, the beautiful strawberry bed looks sparse and the pathways look unruly. In addition, June-bearing plants usually need to be replaced every four years.

Yellow chard and white alyssum make an attractive border for this sidewalk.

Instead, I prefer to plant everbearing strawberries. They produce fewer berries that are often smaller in size, but they don't produce runners, and therefore "stay put" in the landscape. I've found them to be much better for the edible landscape, where I want plants to stay in their place permanently. I plant a variety called Seascape because the berries are almost as large as June-bearing strawberries. We get about 15 pounds of strawberries each year from our front yard strawberry berm.

Edible Landscaping with Nuts

Nut trees such as walnuts or chestnuts make beautiful shade trees and wildlife havens while also producing a nut harvest. For smaller landscapes, hazelnuts and filberts are produced on shrubs.

Herbs and Flowers in the Edible Landscape

Herbs and flowers are among the easiest additions to the edible landscape. By simply replacing an ornamental flower garden with edible herbs and flowers, you can have beauty and function, too.

White cherry tree blossoms stand out against the beautiful dark bark.

Favorite Combinations

- Dill is a cheerful companion to pink cosmos in a flower garden along a fence line.
- Bronze fennel pairs with white cosmos for a more sophisticated look.
- Silvery sage supports pink-flowered ornamentals such as roses.
- Thyme makes the perfect border along a walkway or flowerbed.
- Chives are a beautiful border in the strawberry bed, where they also fertilize and repel pests.
- Oregano pairs beautifully with wild geranium in a pollination garden.

Herbs for Fragrance

Chamomile, lavender, sage, lemon balm, and rose produce wonderful scents alongside a walkway.

Edible flowers are a genius way to add color to your landscape and to your kitchen creations. Anise hyssop, calendula, chives, elderberry, lavender, Johnny jump-ups, nasturtium, and runner beans are some of my favorites.

Beautiful, Deer-Resistant Edible Landscaping

Sadly, there are limited options for those living in deer country. Tall deer fencing negates any aesthetic appeal of edible landscaping. However, there are a few plants that can make beautiful additions to the landscape. While these are known to be deer-resistant, they are not deer-proof, and should be protected when young.

Deer-Resistant perennials

Perennials can add beauty and deliciousness to the edible landscape in deer country, and be low-maintenance, too!

- Asparagus fronds—which grow to six feet tall—make an interesting backdrop to a landscape or can frame a fence line.
- Currants and gooseberries have beautiful leaves and berries, and make great hedge plants.
- Rhubarb—with its beautiful red stalks and giant leaves—makes an interesting addition to the edible landscape.
- Fig trees have gorgeous leaves, and now that there are winter hardy varieties available, almost everyone can grow one. They can also be trained to grow in a pot that is brought inside over the winter.

Tall pink cosmos blend well with yellow dill in a fenceline flower garden.

Flowering sage makes an attractive companion to the rain barrels along this walkway.

Rugosa rose varieties have beautiful, fragrant flowers and large, vitamin-C-rich hips.

When you find a plant—like calendula—that is beautiful, edible, prolific, attracts beneficial insects, and readily self-seeds, you plant it everywhere and then save its seeds!

- Goumi is a shrub that grows six to 10 feet tall and can be sheared to a specific size for compact landscapes. It is a low-maintenance shrub that thrives on neglect. When in bloom, goumi shrubs are pleasantly scented. The sweet-tart, red, cherry-size berries are a nutrient-dense food source.
- Oregon grape-holly is an evergreen shrub that makes a nice hedge and produces purple berries in bunches like grapes that make the most delicious jams and jellies.
- Pawpaw trees are both deer resistant and an excellent way to add native fruit trees to the Eastern North American landscape.

Deer-resistant Herbs

Most herbs tend to be deer resistant, but chives, dill, fennel, lavender, lemon balm, mint, oregano, parsley, rosemary, sage, and thyme are beautiful herbs that work especially well in the deer-resistant, edible landscape.

Deer-resistant Vegetables

Cucumbers, eggplant, peppers, and tomatoes tend to have the most success, but some deer are hungrier than others. Try them and see what works for you! Cucumbers can occupy a well-placed trellis against the house, eggplants work well in a purple-themed garden or with a white-flower backdrop, and peppers and cherry tomatoes work well in a colorful landscape.

Utilizing the Parking Strip

The parking strip—that unused sliver of grass between the street and the sidewalk—is often overlooked as a location for growing edibles. Yet as more and more people are aiming to transform their homesteads into units of production, unused space is getting a second look: What if the parking strip could be both beautified and productive at the same time?

Ron Finley, a radical gardener of unused spaces in Los Angeles, caused a stir a few years ago with his guerrilla parking strip gardens as a way to combat urban food deserts in his neighborhood. Because the local municipality owns parking strips, he was initially reprimanded and ordered to remove the gardens. After some activism and conversation, parking strip gardens were eventually permitted as long as they followed certain guidelines. Though initially an urban statement, parking strips are now being utilized for food production in a variety of residential environments.

The Suburban Strip

The trend of planting productive gardens in unique spaces hasn't skipped the suburbs, however. I recently read an article about farming your parking strip that highlighted a charming suburban residence in Seattle, Washington. Aesthetically built raised beds were built and filled with a tasteful mix of edibles, pollinator gardens, and other features, while allowing easy access between the street and the sidewalk. Imagine the yields our cities could produce, and the increase in our food security, if we simply grew fruit trees in the parking strips and planted nut trees as our shade trees!

This idea of parking strip farming didn't skip my yard, where we added three dwarf cherry trees. In our yard, the parking strip happens to get the most sun, so why not put it to good use? The trees should yield about 30 pounds of cherries annually. By farming the parking strip and reducing the lawn, not only are we increasing our edible yield, but we're also adding beauty, providing habitat for pollinators, and increasing biodiversity.

Planting Considerations for Parking Strips

Planting outside the box brings up some interesting considerations. Here's what to look out for before planting in your parking strip.

Follow your municipality's guidelines for the space.

Most parking strips legally belong to the local municipality, even though property owners are responsible for maintaining them. As such, it will be in your best interest to follow the municipality's guidelines for use of the space. In the city of Seattle—whose guidelines I tend to follow because they are logical and address growing edibles—garden beds in the parking strip must respect pedestrians. Beds must be three feet away from the street and one foot away from the sidewalk. Raised beds should be three feet apart for pedestrians to access the street from the sidewalk, and raised beds should be less than 18 inches tall.

Consider water access.

Before adding plants, notice—does your hose reach? If not, are you up to lugging watering cans? Look for drought-tolerant perennials and Mediterranean herbs that work well in low-water environments.

Avoid dog pee.

It's going to happen, so you might as well plan for it. Fruit will be safe on fruit trees (our choice), but the same can't be said for low-growing vegetables. Consider building a raised bed for protection of vegetables or edible herbs.

Figs ripening on the tree

Cayenne peppers make a cheerful, bold statement in the deer-resistant edible landscape.

Cherry trees make use of the parking strip.

Locate access areas.

Most parking strips have utility pipes and cables running through them. Figure out where they are and leave those areas unplanted. We spaced our cherry trees so there is access to the pipes and cables running underground.

Expect to share.

Parking strips are prime picking spots for pedestrian grazers. Be happy that someone gets to eat fresh, chemical-free produce instead of whatever else they might eat! We haven't had any vandals or secret pickers in our front yard. However, your location might be different.

An edible landscape is a fulfilling project that will increase your yard's productivity, biodiversity, and aesthetic appeal. Edible landscaping techniques can even allow you to grow food incognito in a neighborhood with strict homeowners' association rules.

11

DIGGING DEEPER: PERMACULTURE AND MICRO-FARMING

YOU CAN SOLVE ALL THE WORLD'S PROBLEMS IN A GARDEN.

GEOFF LAWTON

Permaculture is becoming an increasingly popular buzzword to farmers and gardeners. On the suburban micro-farm, you need strategies that help reduce costs and make efficient use of your limited time. In conventional agriculture, time-saving solutions are presented in the form of synthetic fertilizers and pesticides, but often these time-saving techniques come at the expense of the local environment. With permaculture you can rest easy that strategies—when used appropriately—will not only save time, but also enhance the local environment.

Permaculture is basically a toolbox filled with strategies (tools) that help to create a productive and low-maintenance landscape, while improving biodiversity and regenerating soil fertility. The tools in the permaculture toolbox aren't all considered "permaculture" strategies, but they align with permaculture goals. For example, permaculturists didn't create organic gardening methods, but organic gardening is a tool used in permaculture systems because it meets the requirement to nurture the ecosystem.

Modern industrial farming would have you believe that the only way to get a worthy yield in an efficient manner is to use chemical controls at the expense of biodiversity and soil health. Meanwhile, permaculturists and other environmental advocates are demonstrating that when you actively seek ways to regenerate fertility and biodiversity, you can increase the productivity of land and the longevity of the soil. Indeed, as you invite nature onto your micro-farm, you may be able to reduce maintenance time, too.

In this chapter, we'll take a look at some of the tools in the permaculture toolbox that are frequently utilized by permaculturists and may be just the ticket for taking your micro-farm to the next level of productivity and efficiency.

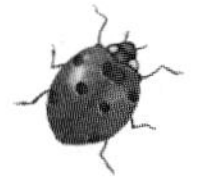 Using Water Wisely

Water is an important resource in any ecosystem. By observing water patterns on a property, you can discover where this resource is being lost and identify appropriate techniques to conserve water and create a landscape that requires little (or no) irrigation, enhances water quality, and results in less work.

In municipal storm water management, water is seen as a nuisance that needs to be whisked away quickly to reduce flooding risk or property damage. Ironically, this whisking away of water often contributes to more flooding and water contamination downstream. How can this be? Before modern civilization began paving over the world one city, parking lot, and road at a time, rainwater infiltrated the earth, where water was slowed, and pollutants were filtered with a great degree of efficiency. That's not to say that flooding didn't happen in the pre-industrial landscape,

but its role in the ecosystem was reduced. Instead of whisking away rainwater, we'll look at some ways to catch it in the landscape.

Water Harvesting with Earthworks

Using earthworks is one way to direct and manage water. Earthworks are earthen, manmade structures that can be used to encourage infiltration, reduce irrigation, control erosion, and enhance soil fertility. Swales and rain gardens are two examples of earthworks often used by permaculturists and other ecological gardeners.

The earth holds a lot of water—especially when compared to impermeable pavement—and only releases water when it is completely saturated. Water is then released in a sort of time-release, draining slowly over time rather than inundating rivers and waterways all at once.

Earthworks are one way to ". . . practice the art of *waterspread*, emphasizing the gentle harvesting, spreading, and infiltrating of water throughout a watershed rather than the rapid shedding or draining of water out of it," according to Brad Lancaster in *Rainwater Harvesting for Drylands and Beyond, Volume 1*. Earthworks help make our remaining permeable surfaces of the cityscape more efficient at catching and holding rainwater.

Catching Water in the Front Yard Landscape

The current popular solution for catching rainwater is to use rain barrels, which are a trademark of the "green" movement. My roof collects over 30,000 gallons of rain per year, a hefty amount of water to take advantage of for irrigating my front yard gardens. However, in many cases, rain barrels are really just an example of over-engineering: They catch an embarrassingly small percentage of total roof water (55 gallons!), the manmade materials need to be replaced over time, and the system creates more work by requiring manual garden watering.

Now, if you have rain barrels, don't worry. I have rain barrels, too. It's important to realize, however, that they have a limited use—such as watering potted plants or a small, nearby garden. When it comes to irrigating a micro-farm landscape, they may not be able to bear the brunt of our expectations. Don't throw those rain barrels out yet! Later in this chapter, I'll share some tools in the permaculture toolbox that can help make rain barrels a more efficient part of a farm system.

For now, let's get back to my front yard landscape. I was partial to using earthworks to capture water from the roof in order to minimize garden maintenance. Mr. Weekend Warrior and I designed a three-element system for catching water. The primary element is a 10-foot-long, one- to two-foot-deep trench that captures the majority of the water coming off the roof. We piled the soil from the trench on the downhill side, creating a berm that absorbs water when the trench

fills up. (The trench and berm together form what is called a swale—a technique explained in the next section.) The berm is densely planted with strawberries, whose three-foot-deep roots assist in absorbing much of the rainwater. From there, excess water is directed into a rain garden, which acts as a fail-safe backup. All three elements—the trench, the berm, and the rain garden—were constructed so that they could each handle the rainwater on their own. Together, the resilient system is almost unbreakable.

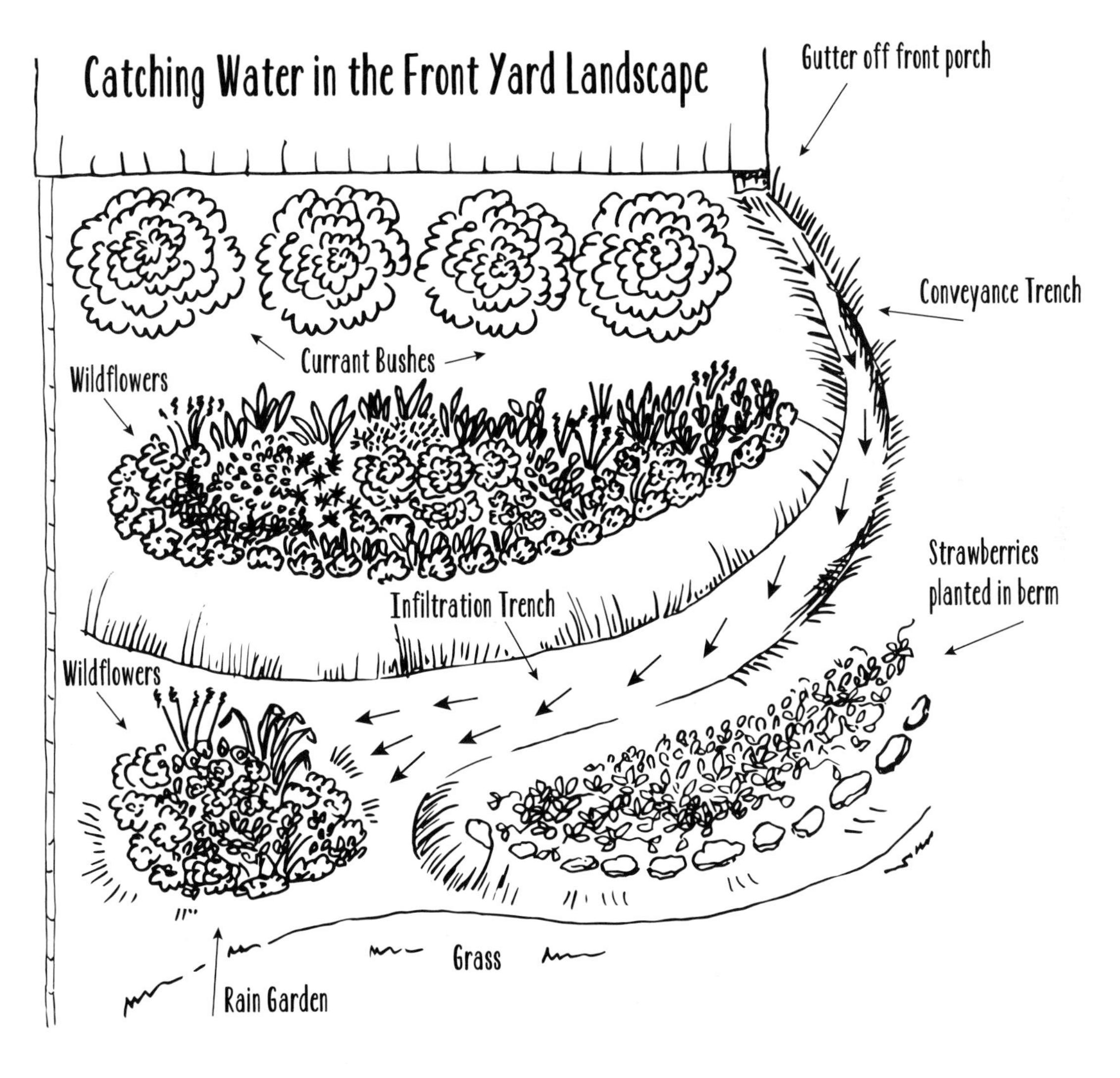

A trench was dug to collect water from the roof. The overflow is directed toward the bowl shaped rain garden. Soil from the trench was piled on the downhill side to form a planting berm and to help absorb rainwater. The trench was later filled with rocks, gravel, and wood chips to double as a walking path.

A conveyance trench directs roof water from the rain chain/downspout to the trench.

Two years after construction. The strawberry berm has a rock border so it resembles a raised bed. Flowers blossoms attract pollinators.

Harvesting Water with Swales

Permaculture swales, often called infiltration swales, bioswales, or berms 'n' basins, are shallow trenches dug on level contours, with a berm on the downhill side. Swales catch water as it drains across flat or gently sloped land, slowing it and spreading it along the contour line.

Imagine a terrace on a gently sloping hillside—it goes across the hillside rather than up and down it. Instead of a terrace wall, however, you see a raised berm for planting, and behind the raised berm is a trench that collects water as it rushes down the hill. This is essentially what we created in our front yard shown in the section above, except we used water from the roof to feed the swale rather than water draining down a hillside.

Swales not only catch water, but also soil, seeds, and organic matter so that they stay on site rather than washing away. As the water is slowed, it infiltrates the ground and creates an underground water envelope, which naturally recharges groundwater and releases moisture during dry seasons. The grass below my front yard swale is consistently the greenest on the street during the dry season because of the water envelope.

Permaculture swales are applicable on both large tracts of open land and at a residential scale.

SWALE PLACEMENT IS IMPORTANT

For maximum efficiency and absorption, swales should be placed as high in the landscape as possible. A swale must be properly located:

- 10 feet away from a building (water must drain away from building)
- 18 feet away from the edge of a steep slope or septic drain field
- Uphill from a low spot that doesn't drain well
- Where an infiltration test demonstrates an infiltration rate greater than 0.5 inches per hour (Go to www.TenthAcreFarm.com/tsmf-companion for a link to instructions to conduct an infiltration test.)

How to Create a Swale

A leveling device will identify a contour line, which is then marked with flags or stakes. The swale trench is dug along the contour line one to three feet deep and one to four feet wide. The earth dug from the trench is piled on the downhill side to make a raised mound or berm roughly one to three feet tall and one to four feet wide.

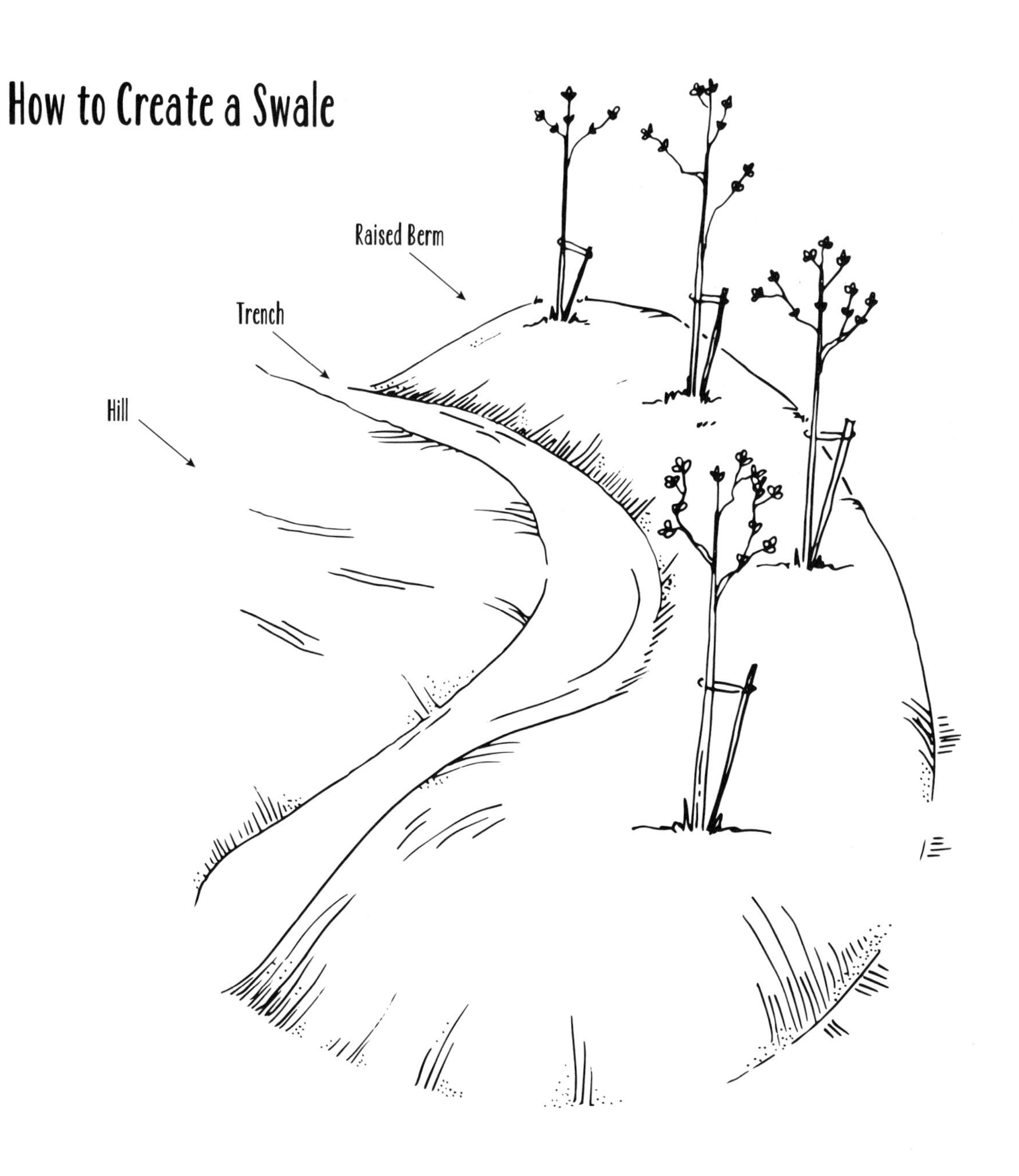
How to Create a Swale
Raised Berm
Trench
Hill

A series of swales can capture rainfall and spread water evenly across a landscape. Multiple swales will be closer together on steep, compacted, clay soils, and farther apart in flat, dry, or sandy soils. A typical distance between swales is 18 to 30 feet. If multiple swales will be constructed, consider how the spaces in between will be used. Decide on an ideal width for the relevant activity, such as lawn mowing, child's play, animal foraging, etc.

Swale berms can be sheet mulched (see chapter 3) and planted with trees and shrubs to hold the berm in place and slow evaporation. In wetter climates, the planting on the berms is for better drainage, and the trench is either mulched or seeded with grass, cover crop, or ground cover. In drier climates, it is the opposite: Trees and shrubs are planted in the trenches where water concentrates.

The moisture in the soil in and around a swale remains long after the spring rains have gone. Because of this, garden beds below a swale will be gravity-fed, and remain lush for many weeks longer than the surrounding area. The moist microclimate will require much less irrigation. Humus will build and absorb even more moisture, storing water in the ground deeper and longer than water spreading across the soil surface. Trees and shrubs planted in the swale system will be naturally irrigated.

Construct one swale at a time. See how the first one performs before adding others. A spillway is an important component of a swale, which is a notch in the berm—as wide as the berm is deep—that directs overflow from one swale to the next. Spillways are filled with rocks or gravel to minimize erosion.

If an area has trees and shrubs already established, dig around them when digging the trench, leaving their roots intact. Retaining the existing vegetation for the absorption capacity of their roots is helpful wherever possible.

CREATING SWALES

For more details on creating swales, see *Rainwater Harvesting in Drylands and Beyond, Volume 2: Water Harvesting Earthworks* by Brad Lancaster. You'll find instructions for marking contour lines, calculations to help determine the appropriate depth and height of your swales, how to manage the overflow, information based on soil types and rainfall estimates, variations of swales for unique situations, and more.

When using swales in a residential area, you may wish to add a few aesthetically pleasing details. Trenches are not a feature we're accustomed to seeing in the residential landscape, and in a high-traffic area, they could even be a hazard. To turn the trench into a walkable pathway, lay

This small, front yard swale berm is inconspicuously bordered with typical garden edging. (Photo by Nicky Schauder of PermaKits.com)

This swale garden was constructed as part of a permaculture school garden in Sterling, VA. The trenches were filled with mulch to double as walking paths. (Photo by Nicky Schauder of PermaKits.com)

Paula wrote to me in January, filled with despair about her desolate front yard. Struck by drought in Texas, the grass had died over the course of a few particularly dry years. Several attempts to reseed it failed:

We discussed a swale as a possible technique to retain the water that drains onto her property from the neighboring yard. Catching every bit of rain during a drought would be necessary for reversing Paula's desert-like conditions. She dug a swale in front of her pecan tree, filling the trench with logs and gravel to help retain moisture. She planted the berm with a perennial ground cover and mulched it well (photo at top of next column).

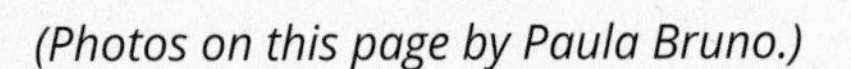

(Photos on this page by Paula Bruno.)

I was surprised to get an update from her in May that looked like this:

This seemingly small swale had breathed life back into her front yard in just a few short months.

down weed blocker fabric on all sides of the trench, then fill with rocks and gravel. The rocks will prevent erosion and moisture loss. Alternatively, top the last two to four inches with wood chips for a more natural walkway. Personally, I prefer the wood chip topping because as the leaves fall in autumn, I don't mind raking up some wood chips, whereas I don't enjoy trying to clear leaves off of gravel. The wood chips will biodegrade and need replacing every two years or so, but in many residential areas they are a free, renewable resource. If your trench isn't in a high-traffic area, you can seed it with buckwheat or clover to prevent erosion and add nitrogen into the soil. Cover with straw until the seeds have germinated.

To improve the aesthetic look of the swale, add a rock border to give the berm the appearance of a raised landscaping bed.

Swales are simply one water-harvesting strategy in the toolbox. While swales catch and spread water across a contour line, rain gardens hold water in a smaller area, which can address our rain barrel quandary. Since rain barrels fill up quickly—one downspout can fill a 55-gallon rain barrel in less than 5 minutes—their usefulness on the micro-farm will be dependent on how you manage the overflow. Sending water back to the sewer system is one option, but it doesn't do much for your garden. Likewise, sending the overflow into the landscape in a non-planned way may mean that you're missing out on taking full advantage of this resource.

Safety is an important reason to think about what to do with the excess water. An overflowing rain barrel can flood a basement, damage a foundation, or worse, do the same to an unsuspecting neighbor downhill.

Harvesting Water with Rain Gardens

A rain garden, also known as an infiltration basin, is a shallow depression in the ground, usually bowl- or kidney-shaped. The depression is most often used to catch water from impermeable surfaces, such as a roof, driveway, parking lot, or even rain barrel overflow. Building a berm on the downhill side of the depression could be useful during heavy rain events. Unlike a swale, rain gardens do not need to be built on contour.

In dryland areas, the depression is a highly efficient area in which to plant trees, shrubs, or other important crops, where the increased moisture and leaf litter will collect, significantly reducing irrigation needs. In areas with more rain, plant the basin with species that can handle soggy root zones for up to 12 hours, such as deep-rooted, drought-resistant, native wildflowers, which will naturally filter out pollutants in the runoff.

When catching water in rain gardens, storm water becomes "an on-site asset rather than a liability," says Lancaster. A rain garden or swale located high in the landscape, such as directly above a garden, is an example of no-work, gravity irrigation.

Always check your local regulations for legality. In my city, rain barrel overflow must legally redirect back to the sewer system, while in the outer-lying areas of my county it is legal to send overflow to a rain garden.

The ideal location for a rain garden is where the ground slopes away from buildings and follows some of the basic safety rules mentioned above for residential swales. Don't send water to within 10 feet of a building, over a septic drain field, near the edge of a steep slope, or into low spots that don't drain well.

Calculate Rain Garden Size for Catching Roof Water

1. Square feet of rooftop x .1 =___
(Volume in cubic feet your garden needs to hold in a 1" rain storm)

Then:
2. Volume / 1.1 = ___
(Surface area in square feet)

Our backyard rain garden, for example, which captures rain from 600-square-feet of roof, should be at least 55 feet squared.

How to Construct a Rain Garden

1. Calculate rain garden size and mark off the area.
2. Dig down 12 inches and grade a flat area across the bowl-shaped surface area.
3. Form a berm on the downslope sides of the garden with the dug-up soil.
4. Loosen and aerate the bottom of the bowl with a digging fork.
5. Add a nine-inch depth of compost soil to the inside of the rain garden bowl.
6. Fill the bowl with water and observe to be sure water will infiltrate within 48 hours. This is essential, because a rain garden that doesn't drain within 48 hours can become a mosquito pond and a human health risk.
7. Plant deep-rooted native plants both in the bowl and on the berm. For a twist, plant the rain garden with edibles.
8. Mulch with shredded leaves or wood chips.

A rain garden can make your rain barrels more useful and eco-friendly. We constructed a long, narrow rain garden to absorb the overflow from two 75-gallon rain barrels in our backyard. Rather

How to Construct a Rain Garden

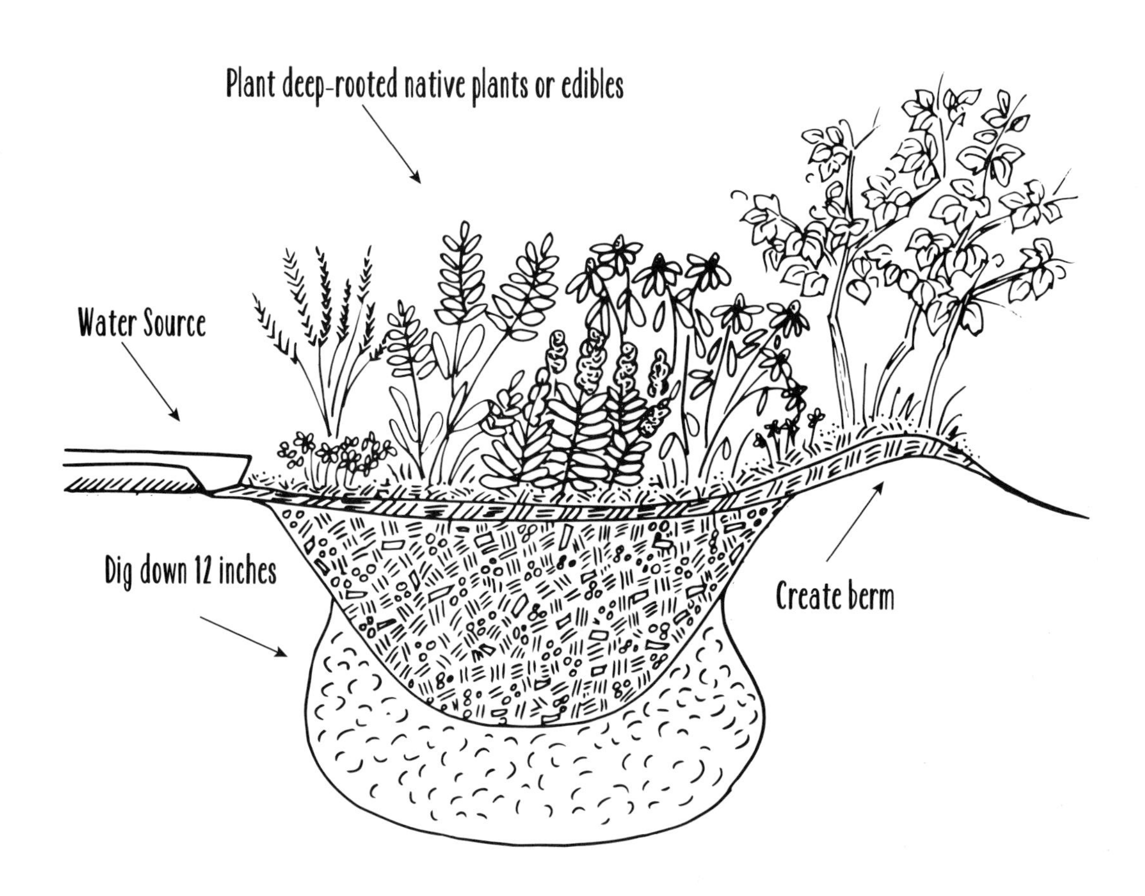

than planting drought-tolerant wildflowers in the depression, we placed flat rocks in the depression to make it a walkable "creek bed." This way, we kept water from the roof in the landscape, while also retaining a pathway from the side yard to the back yard. A berm on the downhill side was planted with drought-tolerant yarrow, rhubarb, and oregano. When the creek bed rain garden fills up, it overflows to a lower-level garden.

EDIBLE CROPS IN RAIN GARDENS

Many edible crops will do surprisingly well in the basin, as long as it properly drains within 12 hours, and is able to dry out in between rains. Some examples of damp-loving plants are asparagus, elderberry, highbush cranberry, red raspberry, rhubarb, spicebush, and strawberry.

If you suspect that the water source might be polluted—such as a parking lot—stick to growing the edible shrubs where the edible part (berries) does not come in contact with the water. According to The Cornell Waste Management Institute of Cornell University, fruits (including fruiting vegetables) are the most suitable for growing in contaminated soils because most toxic metals do not enter the reproductive parts of plants.

Precautions for Swales and Rain Gardens

Water management strategies are site-specific. In some situations, catching water on site is appropriate and useful, while in others it could be devastating. Take note that swales and rain gardens may not be appropriate for all sites. It will be in your best interest to do more research and get advice from a professional before implementing either of these techniques on your property, especially if catching water from a roof.

Construct rain gardens and swales at least 10 feet away from buildings, and be sure water drains away from them. These techniques are not appropriate for areas with high groundwater levels or over septic drain fields. Swales and rain gardens work best on flat or slightly sloping terrain. They should never be located near the edge of a steep slope or on a steep hillside, as this could result in destabilization of the land.

By designing water harvesting earthworks appropriately into your micro-farm design, you can build systems that are long-lasting, efficient, and low-maintenance. They will also recharge groundwater and improve biodiversity.

A walkable "creek bed" rain garden absorbs water from the rain barrels before watering drought-tolerant herbs.

 Building Plant Guilds

A guild is a grouping of plants that supports a central element for maximum harvest and use of space. The use of guilds came about by observing how certain plants would naturally group themselves together in a forest setting, as if to demonstrate that their proximity to one another was mutually beneficial. The concept of designing man-made guilds is relatively new, and many of the early experiments are still in progress. However, guilds provide a way to develop interconnected ecosystems, which may yield more and reduce our workload.

The goal of the guild is to under-plant a central element, such as a fruit or nut tree, with "super plants"—plants that are highly useful and multifunctional. For example, under-plantings of a fruit tree might include plants that fertilize, repel pests, attract beneficial insects, create mulch, and suppress grass.

The general idea is to take advantage of the benefits of plants to reduce costs, labor, and the need to import materials to the farm. Now, to be certain, planting a tree guild will take more effort than simply planting the tree by itself, and it may also cost a bit more at the outset for the extra plants. However, in the long run, guilds will likely be more resilient and vigorous, even if solely from a biodiversity standpoint.

How you use guilds on your micro-farm will depend on your space. Some micro-farms will span several acres, while others will cover less than half an acre. On larger properties there may be space to build a large guild under an expansive, 70-foot tall nut tree, for example, while on smaller properties, the central element will likely be something smaller, such as a dwarf fruit tree or berry bush. If you would like to build a guild, choose a central element that is appropriately sized for your property.

Fruit and nut trees can be linked together in a grouping, under-planting them all with guilds. Toby Hemenway of *Gaia's Garden* would call this a "superguild." The grouping could also be called a *food forest*, which I'll describe in more detail later in this chapter. This grouping of plant guilds could also be created in the shape of a long hedgerow, another tool that I will discuss later in this chapter.

The most common example of a guild is that of the apple tree guild. With an apple tree as the central element, you can prevent grass from creeping under the tree, and repel wildlife, by planting a ring of daffodils and garlic chives at the dripline of the tree. Bee balm, dill, and fennel peppered underneath will attract pollinators. Comfrey, dandelion, yarrow, and white clover will accumulate nutrients and fix nitrogen to fertilize the soil. The comfrey and nasturtiums will provide mulch or green manure. The bee balm, garlic chives, and yarrow will emit strong scents to repel pests. Because apple scab fungus is a common ailment of apple trees, the fennel and garlic chives will provide some anti-fungal properties.

Now, for certain, this is not a recipe, merely an example of how you can take advantage of nature's gifts to create a mini-ecosystem that reduces your workload. Even if you were to follow a so-called recipe such as this one, your guild will likely need tweaking to accommodate the unique conditions of your site. To see how a guild might need to be tweaked, my cherry tree guilds offer an example.

My Cherry Tree Guilds

As I've mentioned before, I planted dwarf tart cherry trees, but I planted them as the central element in fruit tree guilds. Underneath them, I planted a ring of chives around the base of the tree to deter pests. Since the flowers bloom at the same time of year as the cherry trees, they help

to improve pollination (and thereby, fruit set) by attracting more pollinators. The trees are also under-planted with comfrey plants, which are heavy accumulators of essential nutrients, attract pollinators and beneficial insects with their purple flowers, and create an ideal mulch with their giant leaves that can be cut back several times a season.

Everything was going well with my cherry tree guilds until the third year after planting, when I noticed a yellowing of the leaves and insects on the undersides of most of the leaves. I was devastated: it was an all-out breakout. After much research, I finally identified the pest as the hawthorn lace bug. This isn't a common pest to cherry trees, but my trees happened to be located next to a hawthorn tree, and apparently this pest liked my cherry trees, too.

Now, normally our first reaction would be, "How do I kill this pest before it kills my trees?" But in permaculture, we look for a long-term solution. So I did some research and discovered that there is a beneficial insect called a green lacewing that will prey on the hawthorn lace bug. All I would have to do is plant some plants underneath the cherry trees that attract the green lacewing and hope it shows up!

The cherry trees were under-planted with comfrey and chives.

Fruit tree guilds encourage pest-free harvests.

I looked around the yard and noticed the plum tree was not bothered by the hawthorn lace bug. Since cherries and plums belong to the same genus, *Prunus*, I wondered why the plum wasn't affected. I noticed it was under-planted with daisies, and then discovered that daisies will attract the beneficial green lacewing. So I made plans to transplant some daisies under the cherry trees. I also discovered that my cherry tree guilds were lacking in a nitrogen fertilizer, so I committed to seeding the empty spaces underneath the trees with white clover.

Two years later, I got a great harvest (27 pounds) from cherry trees that were free of pests, without having to spray a single thing. So although I started out planting some often-recommended guild plants, in the end, my guilds needed a little something extra that only I was able to figure out because of my unique situation.

The Black Walnut Tree Guild

Black walnut is usually a tree that receives a lot of flack for being allelopathic, meaning that it secretes a toxic substance called juglone to nearby plants, discouraging their growth. It has been challenging for home landscapers to figure out what to plant near them. However, according to *Edible Forest Gardens* by Dave Jacke, *Gaia's Garden* by Toby Hemenway, and *Restoration Agriculture* by Mark Shepard, there are plenty of edible and useful plants that will grow under the walnut tree, evidently unaffected by juglone.

This is great news, because black walnut is a nutrient accumulator of phosphorus and potassium—two important nutrients for plant growth. The nutrients accumulate in the leaves, which will also have accumulated juglone. As the leaves fall, they will fertilize the ground of the juglone-tolerant plants, saving time and money.

Walnuts are an important source of nutrition for humans, and walnut flowers are an important source of pollen for pollinators in the spring. It would have been a shame to not take advantage of this tree!

Fruit Tree Guild Recipe

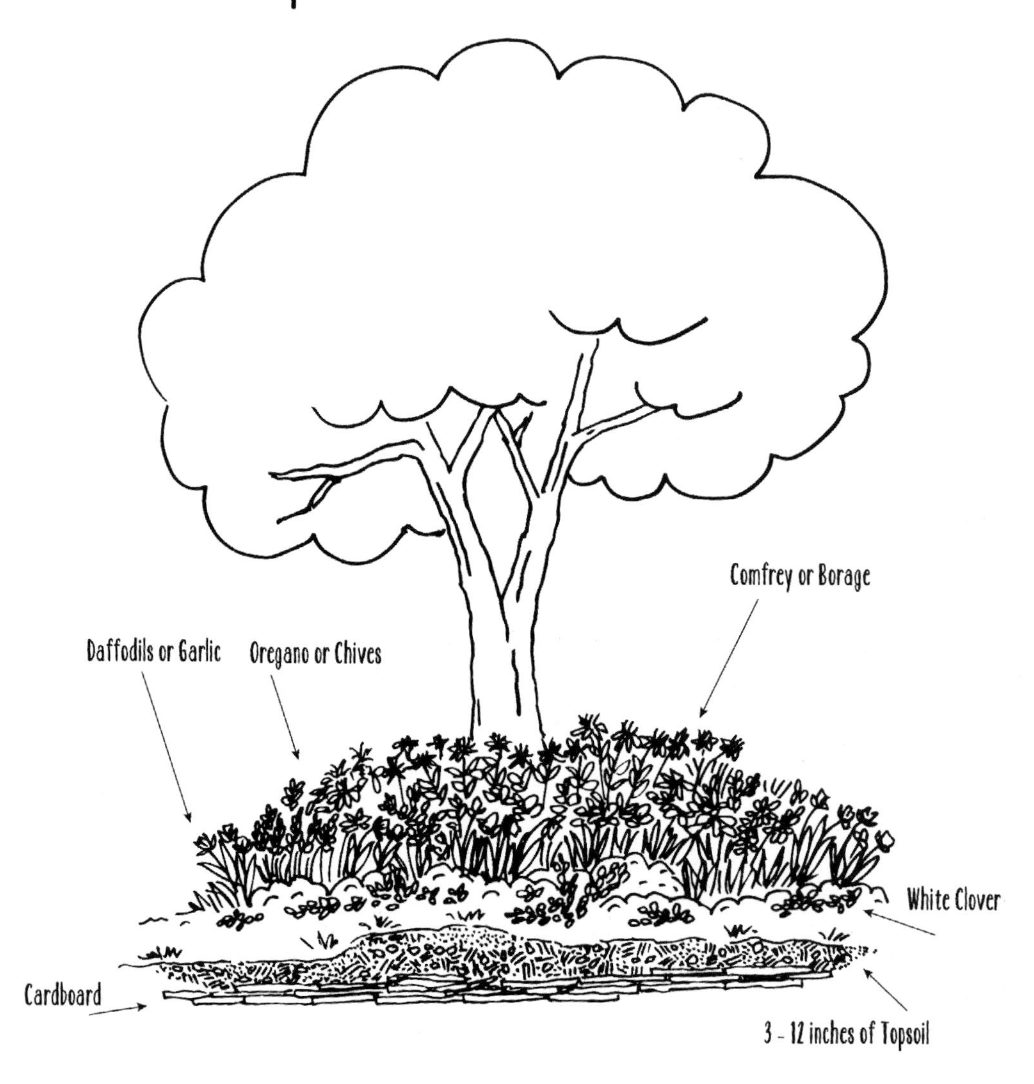

Black walnut is a good example of a larger guild because the tree will eventually reach 50-70 feet tall at maturity. According to the books listed above, cherry, pawpaw, persimmon, plum, and quince are all fruit trees that will grow comfortably underneath the walnut. Farther out from the smaller trees try planting shrubs: black raspberry, currant, elderberry, hazelnut, mulberry, or spicebush. Underneath all of the trees and shrubs can be planted an array of juglone-tolerant herbs for attracting pollinators, deterring pests and wildlife, and fertilizing: alliums, bee balm, dandelion, purple coneflower, white clover, and violet.

The black walnut guild will go through succession, meaning, it will change as the young plants mature, grow taller and fuller, and interact with one another at the root level. Some of the plants may die out as the larger plants cast more shade, while others may struggle with increasing amounts of juglone. This is okay and expected. Rather than expending extra energy to save unhealthy plants, observe and support (and perhaps even plant more of!), the remaining plants that thrive. Focusing on what thrives is an important tool in the permaculture toolbox.

A Fruit Tree Guild Recipe

Permaculture guilds are not exact recipes to follow. They are combinations of plants that people have tried and have observed growing together in natural ecosystems. For example, one day I was hiking in a local park and noticed wild geranium (*Geranium maculatum)* growing densely throughout the forest with wild ginger (*Asarum canadense*), so I planted the two together in a shaded pollinator garden in a corner of my patio where they have thrived. However, just because these combinations were successful in one environment, doesn't mean they'll work in another, so a little experimentation will be in order.

A Rough Beginner's Guide to Guild-making

Guild Step 1

Select a site in full sun with an appropriate amount of space for your tree. Plant your tree according to the instructions in chapter 5.

Guild Step 2

Measure a circle around the fruit tree using sticks or flags to mark the expected mature width. This perimeter is called the drip line. The roots of the tree will eventually extend to this point, and perhaps even farther. Because of this, you will increase success by improving the health of the soil inside this circle.

Spread cardboard under the tree, overlapping the ends so the ground inside the drip line is thoroughly covered. Moistening the cardboard with water is beneficial, but not required. Cover the cardboard with 3 to 12 inches of compost soil, keeping the soil away from the trunk of the tree. Be sure that none of the edges of the cardboard are exposed. It is ideal to allow the soil to rest for at least two weeks and up to three months before proceeding with planting in step three.

Guild Step 3

In this step, you will under-plant the tree with herbaceous plants that: fertilize, mulch, produce nectar to encourage good pollination, attract beneficial insects, repel pests, deter wildlife, and suppress grass. That seems like a lot of plants to fit under one tree! The good news is that some plants perform more than one function ("super plants"). There are a lot of plants to choose from, but the following are some of my favorites:

a. Choose Daffodils or Garlic — Daffodils and garlic repel deer and other wildlife, and repel fruit tree borers. Plant a ring of daffodils or garlic in a circle six inches from the trunk and five inches apart. Daffodils and garlic also suppress grass. To keep grass from creeping into your guild area under the tree, plant a second ring of daffodils/garlic at the outer edge of the drip line.

Oregano is a beautiful and useful under-planting for fruit trees.

b. Choose Comfrey or Borage — Comfrey and borage are both herbs that produce fertilizer, mulch, and nectar, and are excellent at attracting beneficial insects. Plant either Russian comfrey root cuttings or borage from seed. For a dwarf fruit tree, you'll need only two plants per tree, planted on opposite sides of the tree. For a regular-sized fruit tree, plant four plants, one on each side of the tree.

c. Choose Oregano or Chives — Oregano and chives are both aromatic pest confusers, which means that their strong scents will repel pests. They both can also take a

little bit of foot traffic, which will be helpful during harvest time. For a dwarf tree, plant two plants of oregano/chives, and for a regular-sized fruit tree, plant four plants.

d. White Clover — White clover is an excellent source of nitrogen, an essential nutrient for healthy fruit production. It is also often used in orchards as a walkable ground cover. To plant white clover, sprinkle seeds lightly over the entire drip-line area in the spaces between all of the new seedlings.

Water it well and cover the soil with a light mulch such as chemical-free straw. Note: Be sure to only step inside the drip line when it's absolutely necessary for harvesting or pruning. Otherwise, stay outside the drip line to reduce soil compaction.

Now, go forth and create some mini ecosystems (guilds) on your micro-farm!

Building Food Forests

A food forest (See the "Building Food Forests" illustration in this section) is a food production strategy that mimics a woodland ecosystem. In a forest, various plant species grow stacked together in layers to take advantage of the sunlight. Tall trees, small trees, shrubs, herbs, vines, ground covers, and even mushrooms are all intertwined, producing a vibrant, productive, low-maintenance, self-maintaining environment. No one weeds or fertilizes in a forest! A food forest mimics this idea, with the exception that all of the central elements are edible plants and are situated for maximum solar collection.

For example, a food forest might include chestnut trees as a tall canopy tree layer, with apple trees below them as an understory tree layer, followed by currant bushes, a host of edible herbs and mushrooms grown underneath, and perhaps even grapevines that use the apple trees as trellises. An annual pruning can help keep it all in check. This arrangement might triple the yield of a simple apple orchard arrangement in the same amount of space.

In smaller spaces, the tall canopy trees may be omitted in order to conserve space. Food forests work best in areas with sufficient rainfall and in areas that can be managed effectively for the first few years to ensure success of the planting. Try grouping together multiple fruit- and nut-producing tree guilds to build a perennial food forest!

Farming the Contour

Contour gardening is a way to use the land's contours to maximize the use of available resources for abundant harvest yields. Although permaculturists did not invent the concept of contour gardening, it is a useful tool in the permaculture toolbox for sloping lands.

Building Food Forests

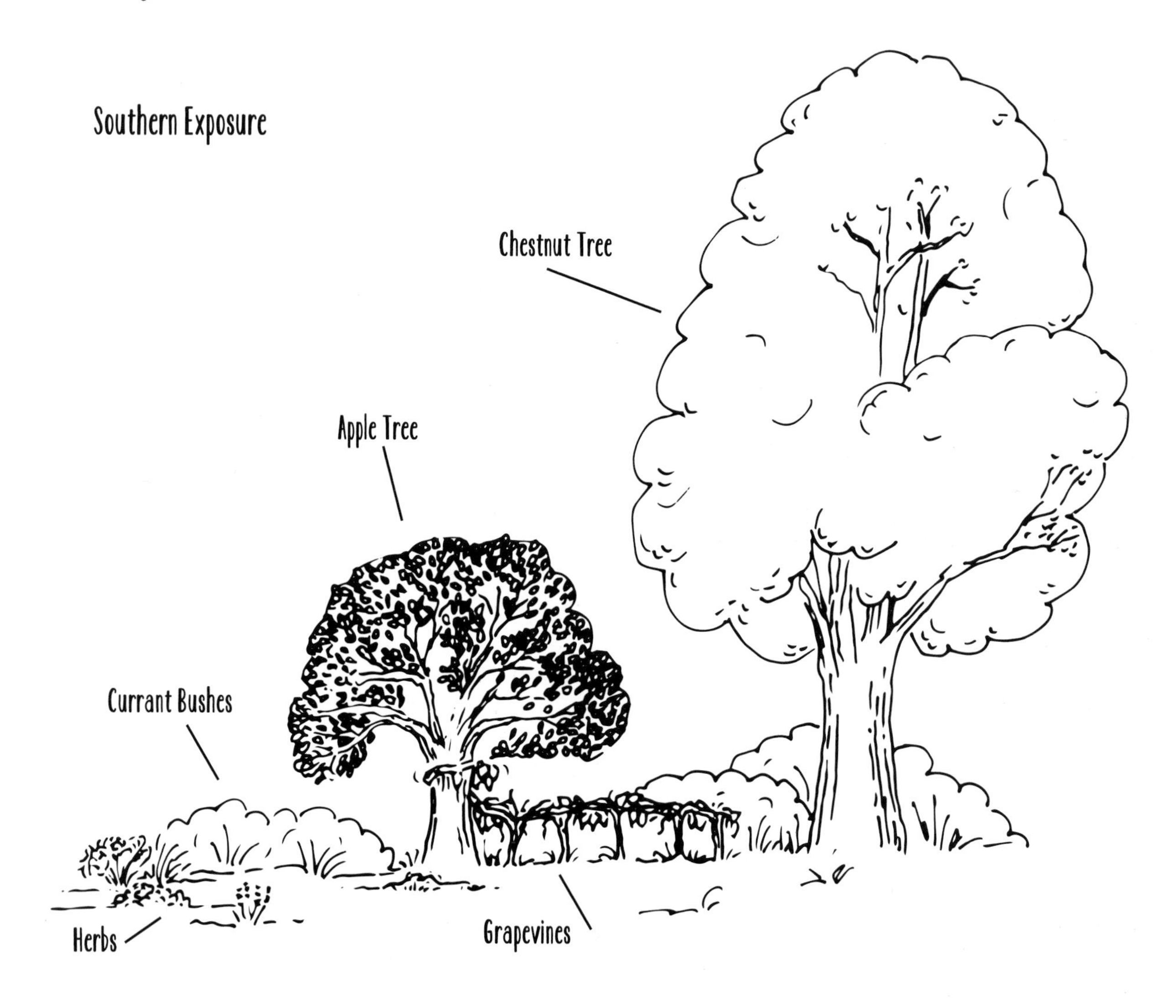

As long as agriculture has been around, farming flat land has been ideal. Gravity, after all, can turn a lovely hillside into a grueling challenge. As population densities increase around the world, however, global agriculture is being pushed from the flat prairie lands to less-than-ideal, sloping terrain. Unfortunately, growing practices like tilling, which were appropriate on flat lands, have been transferred to the slopes.

An inappropriate use of farming techniques the world over has caused an increase in soil erosion and an astounding loss of soil fertility. Contour farming is a cultivation technique that is appropriate for gently sloping lands. In fact, many of the techniques in this chapter help us to work with the land, rather than against it, by using contours to our advantage. Using site-appropriate techniques such as contour farming can mitigate erosion and maximize the absorption of rain and nutrients.

In contour farming, crops are planted across a slope following the natural elevation contour lines, rather than up and down the hill. When planted along the natural curve of the land, gullies and soil erosion are reduced. In this way, water, seeds, and nutrients can more easily infiltrate the soil and be absorbed.

As more and more people are looking for effective solutions for gardening in small and challenging areas, the idea of contour farming has been converted to the backyard scale. In the backyard, raised planting berms built along the contour can be used for either annual or perennial crops. The berms can even be lined with stones or rocks for added erosion prevention and moisture retention. I like the aesthetics of a stone border, too. In a vegetable garden, permanent raised beds can be built along the contour for the same purpose. Contour berms are an alternative to swales, offering an elevated gardening solution for soggy or eroded soil without digging the swale trench.

How to Create a Contour Garden

Contour gardening is a fairly straightforward project. Here are the steps for creating some contour-following planting berms or raised beds.

1. *Mark the elevation contour lines.* Use a leveling tool, such as an A-frame level, to mark the elevation contour lines in the area where you intend to place your garden. Stakes or flags are handy for this. I always keep a stock of them on hand! This will be the center of the garden bed.

2. *Build a raised bed or outline a raised planting berm along the contour line.* Construct raised beds along the contour line that are three to four feet wide. Alternatively, form a berm by first outlining the proposed berm with more stakes or flags.

How to Create a Contour Garden

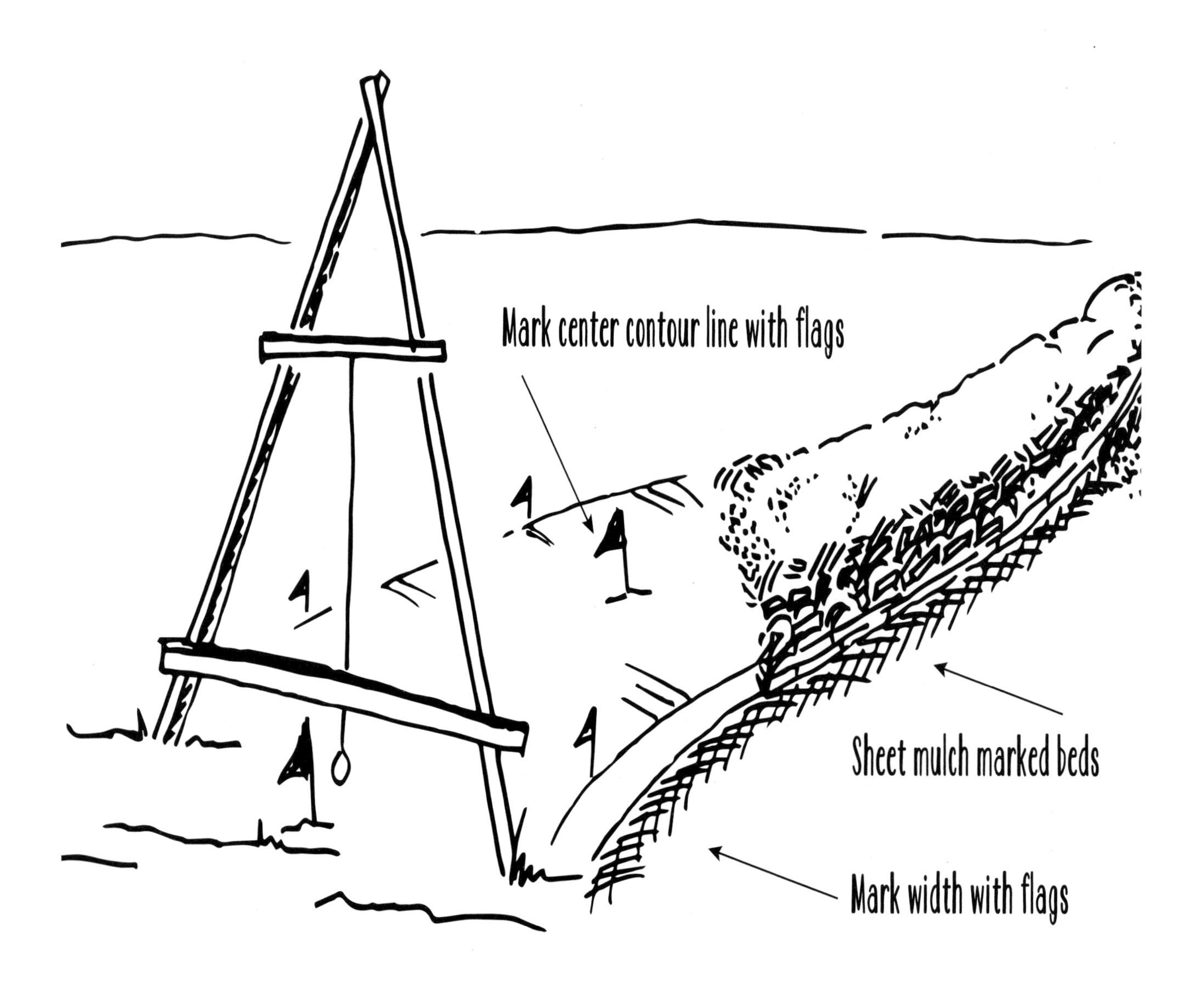

3. *Sheet mulch the proposed garden area.* Use a digging fork to aerate the ground inside the marked bed area, then sheet mulch using cardboard covered with compost soil and other organic matter. (For more details on sheet mulching, see chapter 3.)
4. *Fine-tune the contour garden.* If you have created a berm, line it with stones or rocks if desired to help hold in the soil. Let the new beds rest for at least two weeks before planting; three months is ideal. For this reason, building a contour garden in the fall is a good strategy. During this resting time, expect the soil to settle, so retain extra soil to add to the beds before planting. Now you're ready to plant! This will be an excellent growing area for both perennials and annuals alike.

Contour gardens will reduce irrigation time, mitigate erosion, and retain nutrients on site for a highly productive and efficient garden.

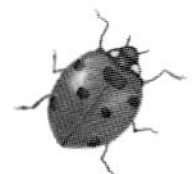 Terracing with Check Logs

Terracing is another strategy that is not a new idea created by permaculturists, but it is one that has been used in many cultures throughout history to grow food on slopes. As more and more of us are challenged with less-than-ideal landscapes, it's important to have a variety of tools in our toolbox to help manage sloping land without increasing soil erosion or further destabilizing sensitive soil.

The subject of "check log" terraces was not something that I was immediately aware of until I began work on a community food forest in 2011. The site was a steep hillside, prone to land sliding and erosion. It was evident that much of the valuable topsoil and nutrients had washed away years before.

We searched for a terracing solution that not only helped stabilize the hillside, but also helped increase fertility (to make up for all that had washed away). In addition, the solution had to be cheap and low tech, because after all, we were just a volunteer group operating on a small budget.

As a permaculture practitioner, I see a gently sloping hillside and generally think, "Swale! Catch, hold, and spread water along the contour!" However, in the case of a hillside that is steep and unstable, the act of digging swales could make a hillside more unstable.

Dave Jacke discusses check log terraces as a solution for this kind of landscape in his book, *Edible Forest Gardens, Volume 2*, while Brad Lancaster discusses a similar strategy he calls check dams in *Rainwater Harvesting for Drylands and Beyond, Volume 2*. We decided to give this technique a try, using their suggestions to create check log planting terraces.

A check log terrace, or check dam, is a permeable barrier that is located perpendicular to the flow of water to reduce runoff and erosion. In simpler terms, what it means is laying logs across the hillside along a contour line, rather than laying them up and down the hill.

Why logs? Logs provide habitat for soil life and critters, and as they decompose they can store more and more water. This technique works well in areas where fallen wood is abundant, and we had plenty of it at our food forest's site at the edge of the woods. In other areas, rocks may be more abundant, and thus a more appropriate building material. It's important to work with existing elements to reduce costs and inputs from off-site.

Benefits of Check Log Terraces

Check log terraces are wonderful assets to the local ecology. Here are some of the ways that check logs build soil and reduce erosion. (See the "Check Log Method" illustration in this section.)

They aerate soil.

As water rushes down a hillside, over time it compacts the soil like a well-used sled-riding hill. In order to have a successful hillside garden, this soil must be loosened and aerated. Remember though, digging or tilling can de-stabilize a steep hillside, so save those techniques for flatter ground. Check logs, on the other hand, are highly aerated, with air exchangeable through the logs and into the new soil behind it. This air exchange in the new soil improves even the old compacted soil underneath it.

They catch nutrients.

Rainwater typically rushes down a hillside, trickles into streams and rivers, and eventually makes its way out to sea. Along this route, nutrients are washed away, never to be available again in the soil. Since there is only a finite amount of nutrients left in the soil, aim to retain as much of them as you can. Check logs are like nets, catching those essential nutrients before they rush away in the rain.

They improve plant life.

Before we built terraces at our community food forest, we tried planting in the existing soil, but the plants looked sad and droopy, and they had yellow leaves. This was an indication of collapsed soil—the air had been pressed out over time as the soil had been compacted. These symptoms also indicate a lack of fertility. With the check logs, plants thrive in the loose, aerated soil, rich in nutrients.

The Check Log Method

Logs and brush are laid across the hillside like a beaver dam and held in place with wooden stakes. New soil is added above the dam and planted with perennial trees, shrubs, and herbs whose roots will soak up and slow the rain as it rushes down the hill. Alternatively, check logs can be built above existing trees and shrubs.

Step 1: Drive in stakes along a contour line every two to six feet. Measure and mark the width of the bed with stakes. We used untreated pine stakes.

Step 2: Lay cardboard as a weed barrier behind the stakes to cover the area wide enough for a garden terrace (usually three to eight feet wide, depending on whether you're planting vegetables or trees). Lay the cardboard like shingles on a roof—start at the bottom of the slope, with top pieces on the top. Overlap the pieces by six inches on all sides—weeds will find any openings. If the check log terrace won't be planted in, and is only being used to slow water, then this step can be skipped.

Lay limbs uphill of stakes to act as a dam.

Step 3: Lay logs, limbs and brush uphill of the stakes to act as a dam. We used logs and limbs that we cleared from the area in preparation for this project. Logs and limbs one to four inches in diameter work best. Logs that are six to eight inches in diameter will work, too, but first, dig a little trench for them to sit in, to take pressure off the stakes. Pile the logs up so that they're slightly higher than level, because the terrace will settle over time.

Step 4: Plug up the holes. The logs and limbs are packed with twigs, brush, and leaf mold to act as a filter, holding in soil and catching seeds, nutrients, and soil during rain events.

Step 5: Add soil uphill of the limbs and brush, on top of the cardboard until it is level with the terrace. The soil will settle over time, so expect to add a bit more in the future. If the check log terrace won't be planted in, then it is not necessary to add soil.

Step 6: Observe the terrace after a couple of hard rains. How did it hold up? Are there any low spots that need more logs, leaf litter, or soil?

Step 7: Plant perennials in the new soil, where their roots will be permanent fixtures in the terrace. Mulch with straw or shredded leaves after planting. Or, for a living mulch, plant clover or

an array of low-growing herbs. If your check log terrace won't be planted in, add extra leaves or straw as the debris settles.

Step 8: Observe over time. As the terrace settles and decomposes over time, you may notice little spots here and there that need to be plugged up, or a stake that needs replaced. In general, though, this little ecosystem you've built will do a good job of stabilizing itself without a whole lot of work on your part.

Pack leaf mold and brush behind the limbs to act as a filter.

Add soil uphill of the limbs and brush on top of cardboard.

Plant perennials in the new soil, where their roots will be permanent fixtures in the terrace.

They increase organic matter.

Rich, aerated soil attracts worms and other beneficial soil organisms. As they go about their daily business, they wiggle in and out of the new and old soil, forming little tunnels everywhere they go. These tunnels are lined with a sticky exudate that they excrete, which helps hold the loose soil together so it doesn't wash away. The tunnels allow air, water, and nutrients to penetrate deeper into the soil, preventing even more runoff. Worms create humus, which improves the health of the soil.

They attract beneficial fungi.

Fungi are an indication of healthy, mature soil, and they show up fairly quickly in check log terraces because of the decomposing logs. Fungi that you see above ground are connected to one another below the soil surface through fungal networks. These fungal networks form beneficial relationships with the roots of the plants, and catch and hold both soil and nutrients.

They build an ecosystem with very little work.

Ultimately, a check log terrace on a slope will become its own self-sustaining ecological system, drastically improving the stability of the hillside and contributing to the regeneration of an eroded landscape.

Check log terraces can help manage sloping land without causing soil erosion. They can also help build soil and improve biodiversity.

Managing the Edges

Managing your edges is an important first step in ecological property design, according to Geoff Lawton, Australian permaculture educator, consultant, and practitioner. By defining the edges, you can better control what comes on your property, such as weeds, pests, wind, aerial chemicals, or water.

The practice of growing hedgerows—a narrow strip of plantings—to define edges stems from at least medieval times in England and Ireland. Ancient hedgerows of the countryside were used as property boundaries, defense barriers, and livestock paddock dividers. These old hedgerows were impenetrable and required a lot of maintenance. After World War II,

skilled laborers were in short supply, and hedgerows largely became unruly. Lack of labor—coupled with the industrial farming boom during which landholders sought to eke out every inch of production from their land—caused hedgerows to largely disappear.

A few discoveries were made as the hedgerows vanished: there was more soil erosion, more pests, more wind, more dust, and far less biodiversity. In areas without heavy tree cover, hedgerows had become essential wildlife corridors. According to the Royal Society for the Protection of Birds, "Hedges may support up to 80% of our woodland birds, 50% of our mammals and 30% of our butterflies. The ditches and banks associated with hedgerows provide habitat for frogs, toads, newts and reptiles."

As a country of pioneers, Americans have become accustomed to wide open spaces. Hedgerows were a habit that never caught on. Since the 1930s, however, there has been some interest in hedgerows due to assistance from the USDA shelterbelt and Agroforestry programs, but it hasn't really taken hold as a standard practice.

Hedgerows are commonly used along a fence line, to mark a property line, or along the foundation of a house. Hedgerows can act as linear guilds (read more about guilds earlier in this chapter) or linear wildlife corridors.

Reasons to Plant a Hedgerow

Suburbanites are always looking for creative ways to mark off the boundaries of their properties. That's because it's rare for side-by-side neighbors to share the same philosophies on pets, children, privacy, and lawn care. Hedgerows are an exceptional way to mark boundaries and create privacy. They can also increase the beauty, productivity, and biodiversity of a property. Here are some of the many reasons your micro-farm might benefit from hedgerows.

Beauty

Hedgerows can be an aesthetically pleasing addition to the landscape. With a diversity of flowering and fruiting plants, what's not to love?

Water Conservation

Hedgerows conserve water by blocking drying summer winds that accelerate evaporation. Did you know that more moisture is lost through evaporation on a cloudy, windy day than on a still, hot, and sunny day? Wind is a game changer.

Hedgerow trees and shrubs will catch and store water in their root systems, especially if they are planted on contour, which is one of the reasons why crops near hedgerows tend to be greener. This means that hedgerows are a great way to reduce your irrigation time in the garden.

Privacy Screen

Though it takes four to eight years for hedgerows to become established, if designed properly for the site, they will eventually fill in the space and provide a nice privacy screen. There's something different about being enclosed by a living fence of plants and trees when compared to a privacy fence.

Food Production

Hedgerows can provide food for humans or wildlife. A food-producing hedge is often called a "fedge." A fedge mimics the diversity usually found at a forest's edge, where prairie and forest converge.

Noise Reduction

Hedgerows can help buffer sound such as the white noise from a nearby busy road. A hedgerow should be planted as close to the source of noise as possible. Be aware the hedgerow will be at its most useful for this purpose when the trees and shrubs have reached their full size.

Windbreak

Strong wind disturbs pollination efforts and stresses plants, thereby reducing crop yields. In windy areas, plants will put more energy into growing strong stalks and branches and will have less energy to devote to flower or fruit production. When a hedgerow is planted perpendicular to the prevailing winds, it can reduce wind speeds by up to 75% at distances up to 10 times the height of the hedgerow on flat land, according to Jude Hobbs, an agroecologist, permaculturist, and hedgerow specialist.

Buffering the wind allows you to create a calm inner environment that is comfortable for entertaining, sitting, or growing healthy crops. It can even reduce heating costs by up to 40%. Place trees and bushes with fragrant flowers in the hedgerow, and you can help mask foul odors from nearby industry or livestock operations.

In flat areas where wind can reach higher speeds, a windbreak can serve as a barrier to filter the air from dust particles and aerial chemical drift. A windbreak hedgerow can also minimize

or prevent snowdrifts, and with proper placement, could reduce snow shoveling time on your driveway after winter storms.

Soil Stabilization

Hedgerows are densely planted with a mixed species of plants that have various types of roots that work together to stabilize the soil. Water will be slowed down as it runs through the hedgerow, which will help reduce soil erosion. (Eroded topsoil is America's #1 export!)

Wildlife Corridor

Hedgerows are linear nature preserves, providing much needed nesting, forage, and shelter for mammals, birds, reptiles, and amphibians. Hedgerows perform an important function of restoring habitat that is often missing in suburban subdivisions where the land was stripped of most of its trees, shrubs, and natural bodies of water for development.

One of the most common habitats to disappear is that of the edge where two ecosystems meet, such as where forest meets prairie or where prairie meets a stream. Habitat edges tend to be wild, weedy, and unruly. At the edge of two ecosystems, however, is where you'll find the most diversity of plants and animals.

Hedgerows mimic this essential edge habitat, but can be designed to pass muster for suburban landscape aesthetics. While your hedgerows are getting established, you might consider adding birdhouses, bird feeders, and birdbaths to begin attracting new residents.

Beneficial Insects and Pollinators

Hedgerows can support a diversity of insect species. If you'd like to see more beneficial insects patrolling your garden or more pollinators coming in for a visit, a hedgerow can do more than a wildflower planting all by itself.

That's because hedgerows consist of trees, shrubs, and ground covers in addition to herbs and wildflowers, all of which flower and fruit at different times and provide a variety of options for pollen, nectar, and shelter. More permanent leaf litter will increase habitat for important insects, and more insects may increase the bird and bat populations. Butterflies will also be attracted to hedgerows for protection. If increasing biodiversity is important to you, a hedgerow will catapult your efforts.

Riparian Zone Buffer

Riparian zones are the land areas along bodies of freshwater such as creeks, ponds, lakes, and rivers. They include the floodplain zones as well as the sloped banks of the waterway. Riparian zones are home to many (endangered) species of wildlife and are also essential for filtering out soil particles, organic matter, agricultural chemicals, and other man-made pollutants before rainwater collects in these bodies of freshwater.

How to Plant a Hedgerow

The type of hedgerow you plant will depend on what purpose you want it to serve, the sun exposure of the area, soil conditions (wet/dry, compacted/healthy, etc.), and susceptibility to deer or other animals. Hedgerows are comprised mostly of perennial species, which are slow growing. Annual species can fill the gaps while a young hedgerow becomes established. For hedgerows to achieve the effects you desire, it will be important to maintain it for at least two years by watering and weeding while it becomes established.

Hedgerows are often used along property lines, but you can also use them to divide sections of a property for animal paddocks or to separate play areas from garden areas. Hedgerows can also be utilized to manage water flow, if they are built in contour beds or swale berms along contour lines.

Hedgerows should be twice as long as they are wide. Dave Jacke, author of *Edible Forest Gardens*, suggests 40 feet as the critical width for creating a biologically rich and fertile ecosystem that attracts and holds beneficial insects, wildlife, and a diversity of plants. In this example, the minimum size of a hedgerow would be 40-feet-wide by 80-feet long.

If your suburban property doesn't allow for a 40-foot-wide hedgerow, plan for it to be at least 10 to 20 feet wide. It will need to incorporate at least a few rows of plants to be effective. By comparison, large-scale farmland hedgerows can be as wide as 100 feet, stretching a minimum of 60 feet from a water source in riparian zones.

Serviceberry bushes bordered with evergreen trees for beauty and privacy (photo by Distant Hill Gardens on Flickr)

Serviceberries are a nice substitute for blueberries when a site doesn't have the acidic soil that blueberries require.

Hawthorn tree blooms buzz with spring pollinators.

Red currant bush

How to Plant a Hedgerow

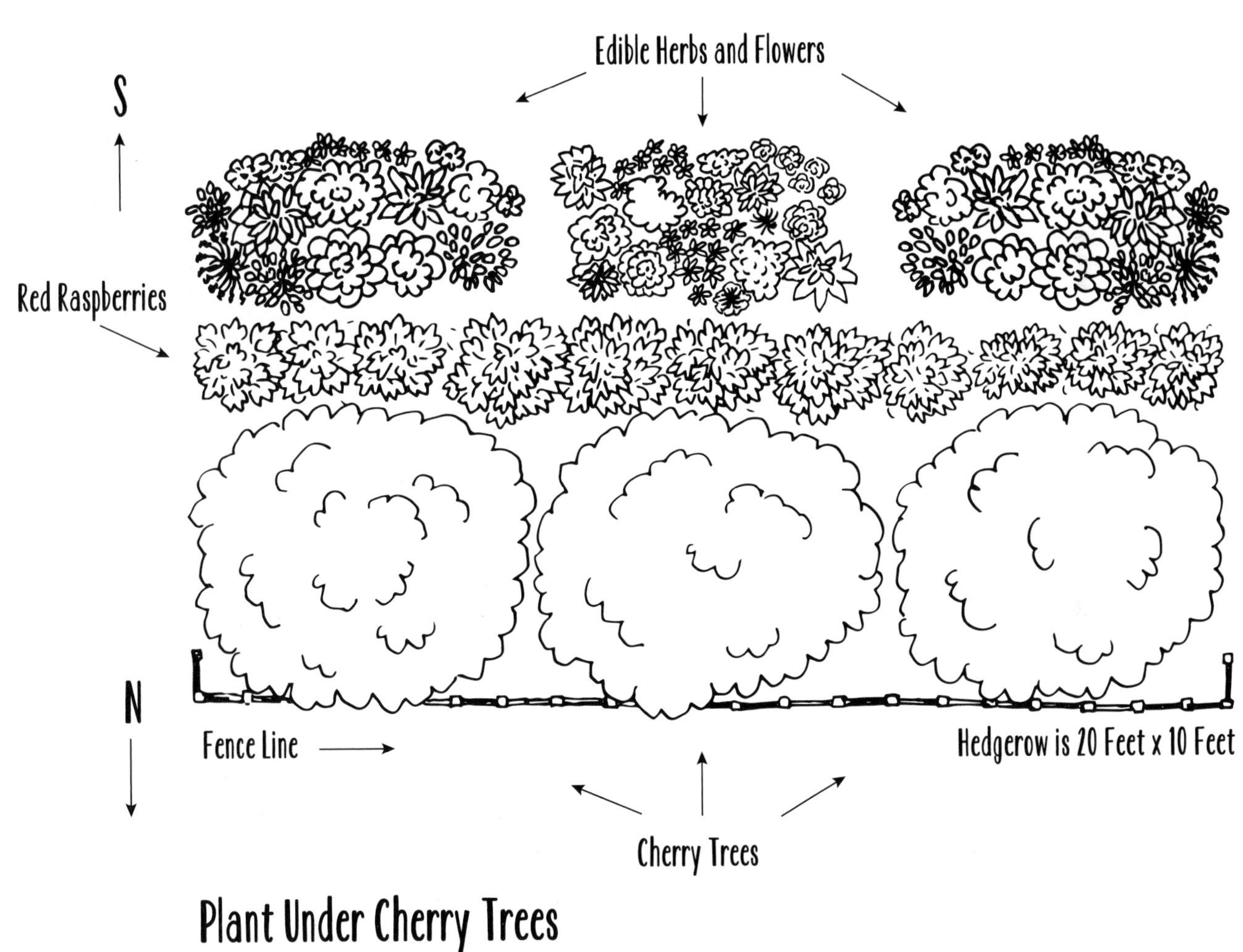

Plant Under Cherry Trees

Daffodils
Clover
Comfrey
Borage
Chives

How to Plant a Hedgerow

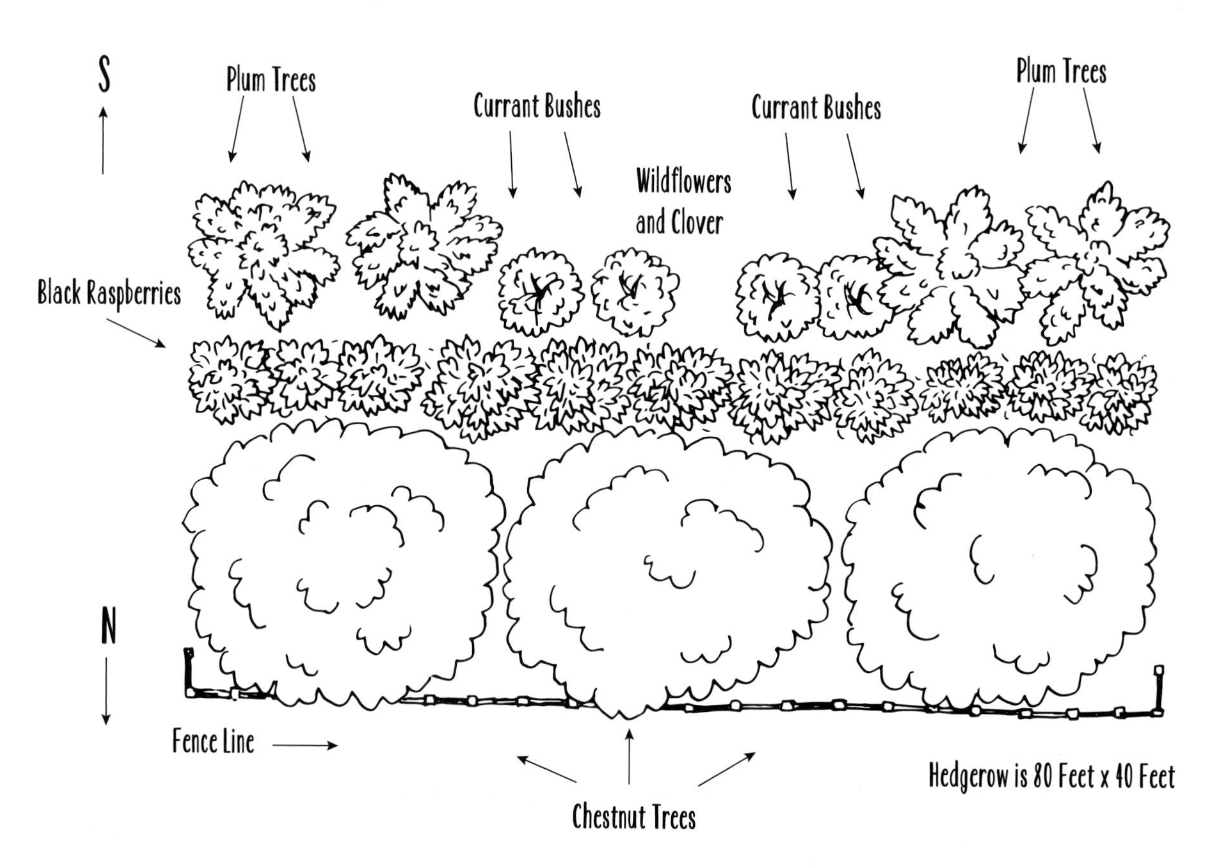

Plant Under Young Chestnut Trees

(While sunny underneath)

Daffodils
Lemon Balm
Anise Hyssop
Comfrey
Yarrow
Bee Balm
Butterfly Weed
Goumi
Fennel

Plant Under Mature Chestnut Trees

(Once growth produces shade)

Daffodils
Lemon Balm
Comfrey
Oregano
Ramps
Chives
Mint
Goumi
Fern

Perennial Maximilian sunflowers—tall wildflowers that attract pollinators and wildlife—peek through the garden fence.

California poppies will grow thickly from seed, covering the ground and attracting pollinators until your hedgerow grows up and shades the soil.

Goumi is a good shrub for a deer-resistant or edible hedge. Beautiful with delicious berries, it will fix nitrogen in the soil, too. Because it will lose its leaves in the winter, it may be planted in front of an evergreen species to maintain privacy.

(Photo by Tatters on Flickr)

Figs can be enjoyed in a deer resistant hedge.

To prepare the area before planting, sheet mulching is a good option. To sheet mulch the area, cut back any unwanted growth and remove unwanted woody plants. Use a digging fork to aerate the soil throughout. Cover the area in cardboard, overlapping the ends so that the soil is entirely covered. Amend hardpan soil with a layer of aged manure topped with shredded leaves or straw, then top with at least four inches of compost soil. Wait at least two weeks before planting. On large-scale properties, it might be necessary to till the hedgerow area rather than sheet mulch. Add two inches of compost and let rest before planting.

Your suburban hedgerow can include a variety of canopy and understory trees, fruit trees, berry and nut bushes, flowering and native trees and shrubs, evergreen trees and bushes, and herbs, flowers, and ground covers.

Choose the Foundational Plantings

The tallest plantings will establish the foundation of the hedgerow. In a suburban setting with a 20-foot-wide-x-40-foot-long hedgerow, you'll likely skip tall canopy trees and start with semi-dwarf trees or shrubs as your base planting.

Organize your hedge plantings so the tallest species are closest to the property line or fence. Work inward to layer your plants from tallest to shortest. Give each plant 75% of its suggested plant spacing to ensure that the hedge is full and compact at maturity. For example, dwarf apple trees are expected to get eight to 10 feet wide. To plant a row of apple trees in a hedge, plant them about 3½ feet away from the property line or fence, spacing the trees about seven feet from one another. Alternating them with evergreen bushes might help with privacy, if that is a concern.

Mahonia, also called Oregon grape holly, is an evergreen shrub that reaches around four to eight feet wide. It is often used in privacy and wildlife hedgerows, although the purple berries are edible for humans, too, and make a delicious jelly. Plant *Mahonia* about three feet away from the property line and six feet from each other to accommodate the maximum width.

You may weave a walking path into the design so you can easily visit and maintain the area without stepping on plants or compacting soft garden soil.

Choose the Support Plantings

Shrubs, herbs, flowers, and ground covers all make up the support species of your hedgerow. Your second tallest plants will be planted slightly in front of—and staggered in between—your foundation plants. Support plantings facing south or west will receive more sunlight over time than plants facing north or east, so choose appropriate plants for the sun exposure. To plant hazelnut shrubs that are expected to reach 10 feet wide, for example, plant them about five feet away from the base of the apple trees.

Continue adding species according to diminishing height and width. Red currants expected to reach five feet wide would be staggered about six feet in front of the hazelnuts, for example.

Shade-tolerant perennial herbs can be planted underneath the trees and shrubs, or try growing mushrooms in shady spots. Sun-loving wildflower seeds and clover can be sprinkled throughout to fill in the spaces until everything is established.

The plants you choose will depend on the function and location of your hedgerow. For most functions, a mixture of trees, shrubs, nitrogen fixing plants (for fertilizer), and herbs will create the most successful hedgerow. To create a self-sustaining ecosystem, choose plants that perform more than one function. For example, a holly bush can be a windbreak, privacy screen, and bird habitat. Yarrow will attract pollinators and beneficial insects, help break up clay soil, and accumulate nutrients for fertilizer.

The "Some Plants to Choose for Hedgerows" table in this chapter lists various plant species that will do well in a hedgerow, but this isn't an exhaustive list, and your hedgerow need not be limited to these suggestions. You may need to do more research to find plants that are appropriate to your climate.

Nitrogen Fixers in Hedgerows

Nitrogen is an essential nutrient for self-maintaining ecosystems. Nitrogen-fixing plants convert nitrogen in the air into a usable form in the soil for the plants that surround them. It is recommended that up to 50% of your plantings be nitrogen fixers to naturally fertilize. So if you choose three trees, three shrubs, and three herbs, then consider interspersing up to 9 nitrogen fixers throughout the area, as well.

Deer-Resistant Plants in Hedgerows (NOT deer proof! Protect while young!)

A deer fedge is an edible hedge planted on the outside with food for wildlife, and planted on the inside with food for humans. When densely planted, the deer might prefer to stay on the outside and munch on things rather than jump into your garden! You will need a decent amount of space to create a deer fedge. It might be six feet wide on the outside and six feet wide on the human side by 24 feet long. Densely plant shrubs and trees that are of varying heights in the fedge. This will deter the deer and also serve to confuse their limited depth perception.

Hedgerows will require some maintenance in the first one to four years. Water your plants in dry periods, plant in a rain-harvesting swale, or install an irrigation system. Continue to mulch annually to maintain control over the weeds. After the system is established at full size, the hedgerow should be a self-maintaining ecosystem that requires very little maintenance, save a once-per-year pruning and—hopefully—lots of harvests.

SOME PLANTS TO CHOOSE FOR HEDGEROWS			
Function	Trees	Shrubs	Herbs
Deer-Resistant	American holly birch buckeye cornelian cherry fig mimosa pawpaw pine	barberry boxwood cypress currants gooseberry goumi highbush cranberry holly bushes juniper leatherleaf mahonia red elderberry staghorn sumac viburnum	anise hyssop calendula California poppy catmint daffodil fern lavender lemon balm marjoram oregano ramps rosemary sage sweet alyssum sweet woodruff tarragon thyme
Edible	American persimmon apple cherry cornelian cherry crabapple American hawthorn hazelnut mulberry pawpaw peach pear plum	blueberry currant elderberry gooseberry goumi Nanking cherry rose serviceberry staghorn sumac	anise hyssop calendula California poppy catmint fern lavender lemon balm marjoram oregano ramps rosemary sage sweet alyssum sweet woodruff tarragon thyme

SOME PLANTS TO CHOOSE FOR HEDGEROWS			
Function	Trees	Shrubs	Herbs
Flowering (scent/beauty/ pollination)	flowering dogwood red osier dogwood	false indigo lilac Maryland senna witch hazel	anise hyssop calendula clover dandelion fennel Russian comfrey yarrow
Nutrient Accumulator (fertilizer & mulch)	apple black locust flowering dogwood walnut		clover dandelion lupine Russian comfrey yarrow
Nitrogen Fixer (fertilizer)	alder black locust	false indigo goumi Maryland senna	crimson clover Dutch white clover lead plant round-headed bush clover white prairie clover yellow bush lupine
Privacy/Noise Reduction	sugar maple American chestnut	boxwoods holly bushes junipers mahonia yews	
Tolerant to Moisture (for wet and erosion prone areas)	flowering dogwood pussy willow	elderberry gooseberry hazelnut highbush cranberry lilac staghorn sumac	cattail Maximilian sunflower miscanthus grasses (native, non-spreading)

SOME PLANTS TO CHOOSE FOR HEDGEROWS			
Function	Trees	Shrubs	Herbs
Shade Tolerant	flowering dogwood	currant elderberry gooseberry hazelnut serviceberry spicebush staghorn sumac witch hazel	chives fern lemon balm marjoram mint oregano ramps parsley
Wildlife	crabapple cherry dogwood hawthorn mulberry	American cranberry blackberry/raspberry blueberry holly serviceberry spicebush	aster bee balm butterfly weed clematis coneflower
Windbreak	black locust hackberry oak poplar spruce walnut	American cranberry elderberry hazelnut lilac	

TIP: PLANTING HEDGEROWS

You don't have to plant an entire hedgerow at once. Plant it in sections or in layers—all tall trees first, or one length of fence at a time, for example—so you're not overwhelmed or overburdened by the price of perennial plants. Hedgerows are best planted in the spring or fall. In hot weather, plant perennials on a cloudy day. After planting, water and mulch well to reduce weeds.

Permaculture Plants

Plants are amazing assets to your farming pursuits. Using the right plants can help reduce soil erosion and build soil, attract beneficial insects and repel pests, fertilize, and much more. Here are some of my favorite plants and a few ways to use them for reducing your workload and garden costs.

Protecting Soil with Spring Ephemerals

Many spring ephemerals, such as daffodils, dazzle the early season with their colorful flowers. But their benefit to the ecosystem extends beyond beauty. Ephemeral means "lasting for a very short time," and spring ephemerals are herbaceous plants that leaf out, gather energy, flower, and reproduce in the early spring before the trees leaf out. They can help protect and enrich soils, and increase biodiversity. Because they are perennials, they'll come back each year without any work on your part. Another advantage of spring ephemerals is that they are largely resistant to deer and other animals.

During their short time of spring activity, they have access to full sun conditions while other plants are still dormant. This period fuels their energy needs for the entire year. They reproduce by setting seed as well as by sending out new shoots from the root bulbs, which is why they usually grow in patches or colonies.

Once the trees and shrubs surrounding them begin leafing out and temperatures start to warm up, spring ephemerals are signaled that it is time to go dormant until the following year. The leaves and flowers will die back naturally for the season.

Spring ephemerals contain a high level of nutrients. That's because the early spring rains are rich in nutrients. As the spring rains wash over dormant fields, forests, and gardens, they take topsoil and organic matter with them. Spring ephemerals, with their root systems being active during this time, catch and hold the moisture and important nutrients and prevent them from leaching away. When spring ephemerals die back at the end of the spring season, the excess nutrients they collected will then enrich the plants and trees that surround them.

For this reason, spring ephemerals are an important component of a soil erosion mitigation plan in deciduous forest areas, and you would do well to incorporate them into your micro-farm. Plant them high in the landscape above your garden at the edge to catch and slow down the water. They can even be planted on a slope as part of a hillside stabilization plan.

The flowers of spring ephemerals will provide nectar for early season pollinators, which in turn can improve the pollination of fruit trees for better fruit set. That's one reason why daffodils—a

popular and well-known spring ephemeral—are often planted in fruit tree guilds. However, there are plenty more spring ephemerals that can be beneficial to our micro-farm ecosystems. Here are some spring ephemerals to try.

Ramp (Allium tricoccum) USDA growing zones 4-8

Ramp, also called wild leek, is probably the most useful spring ephemeral because it is deliciously edible. Dave Jacke, author of *Edible Forest Gardens*, calls it, "the king of the edible spring ephemerals." It is one of the most reliable and easiest to grow spring ephemerals in full shade conditions. Just be sure it has access to the spring sunshine.

Since ramps do best in shady conditions, you'll have to wait for your garden to mature and cast consistent shade before planting. Both the leaves and the roots of ramps are edible, similar to their cousins—onions, leeks, and garlic. To ensure that a colony of ramps survives, it's important to leave most of the root bulbs intact. Ramps put out only two leaves each spring. Harvesting only one leaf of each plant will ensure they can still collect energy from the sun, store it in their roots, and produce flowers.

Daffodil (Narcissus spp.) USDA growing zones 3-10

Daffodils are a spring ephemeral that has been widely cultivated along with other spring ephemeral flowers such as snowdrop, hyacinth, and crocus. These dependable spring ephemerals have leapt out of the forest and into sunny yards without skipping a beat. They are an extremely useful, non-edible plant for the permaculture garden. Toby Hemenway, author of *Gaia's Garden*, suggests using them for grass suppression. When planted in a circle at the drip line of a fruit or nut tree, or around the boundary of a garden, daffodils keep grasses from creeping in, preventing the grass from competing with the tree or garden crops for water and soil nutrients.

Daffodils are a Spring ephemeral.

The large bulbs will repel digging animals such as gophers, while the aboveground leaves will

repel deer and other browsers. For that reason, it is common to see another circle of daffodils planted directly around the trunk of the tree.

Camas (Camassia spp.) USDA growing zones 3-8

Camas are another spring ephemeral with a beautiful flower. Similar to daffodils, they have evolved to be useful in full sun conditions, and can be used to suppress grass.

According to *Gaia's Garden*, "camas bulbs were a principal food of western Native Americans and are having a resurgence among wild-food enthusiasts." In order to eat the bulbs, the whole plant must be harvested, so they are best used in gardens where the bulbs can be replaced each year. Never use camas and daffodils together if you intend to eat the camas, since daffodils are toxic and you wouldn't want to get the two confused during harvest.

Camas in bloom.
(Photo by Tom Brandt via Flickr)

Spring Beauty (Claytonia) USDA growing zones 3-8

Spring beauty is a dainty, rose-colored flower that is an important early-season nectary for pollinators. According to Dave Jacke in *Edible Forest Gardens*, it may also be juglone-tolerant, and able to exist under walnut trees. Jacke suggests planting spring beauties with strawberries in partial shade. Just as the spring beauties go dormant, the strawberries will begin leafing out.

Toothwort (Dentaria) USDA growing zones 4-7

Toothwort prefers to grow in the rich, moist soils of deep forest. In fact, toothwort can tolerate wet feet. It makes a nice ground cover, though it will die back in the heat of summer. The leaves are edible (raw or cooked) and nutritious, and are often added to salads or soups.

Never harvest more than 1/3 of a stand of spring ephemerals.

For the spring ephemerals that will be harvested from a forest, it is important to follow the forager's rule of thumb: Never harvest more than 1/3 of any colony of a spring ephemeral. They are

slow growing, and harvesting too much of them at one time significantly reduces their ability to sustain or build a colony. Reverence is key in the forest.

Never collect seeds or plants from the forest.

It is important that we leave our remaining forests as intact as possible. The spring ephemeral colonies (with the exception of daffodils) are fragile and must be protected. If you intend to grow any of these plants in your food forest, it will be easy to source them from native plant nurseries, either locally or online.

When your goal is to reduce soil erosion and increase early spring flowers for pollinators, spring ephemerals may be just the ticket.

Utilizing Super Plants

Seeking out plants that serve multiple functions increases biodiversity, thereby increasing the self-sustaining nature of a micro-farm ecosystem. With the utilization of multi-functional plants, you can spend less time mulching, dealing with pests, and fertilizing. I call these plants *Super Plants*. Here are some of the benefits of super plants.

- **Protect:** Many super plants are fast growing, so they can quickly cover bare ground to protect it. Their roots hold soil together and keep it from eroding away in the wind or rain.
- **Fertilize:** Many super plants accumulate vital nutrients from the subsoil and bring the nutrients into their leaves. As their leaves die back, they make a healing medicine (fertilizer) for gardens and damaged topsoil.
- **Condition:** Decaying roots add organic matter to the soil, provide channels for rain and air to penetrate, and create tunnels for worms and other beneficial soil microbes.
- **Attract:** Many super plants are quick to sprout, but relatively short-lived. For this reason, they flower frequently in order to set seed for the next generation. The flowering and their dense foliage can attract beneficial insects looking for habitat or nectar.

Many super plants will actually increase the productivity of your garden if you know how to harness their power. The following are some of my favorite super plants. You might be surprised to find out that many of them are considered weeds! Plant them around perennial crops, in guilds or hedgerows, at the edge of the vegetable garden, and in the herb garden, too.

Chamomile *(Chamaemelum nobile)*

Not only is the dainty chamomile flower cute as a button, but it works hard in the garden as well. Chamomile has been called "the plant's physician" because it supports and appears to heal almost any plant it is planted next to. Chamomile is a fertilizer plant; its roots dredge up potassium, phosphorus, and calcium, so mulching with the plants will help improve soil. The flowers attract pollinators, and beneficial insects are attracted to the lacy foliage. It is said to especially improve cabbage and onion crops, and it works well under fruit trees, too.

You may know chamomile best as an excellent tea with calming properties. Now's your chance to grow your own!

Chives *(Allium schoenoprasum)*

Chives are a more common herb, and for good reason. They are useful in the kitchen and easy to grow. I love to walk outside in the middle of cooking and quickly snip a few leaves. The flowers are gorgeous and make a delicious edible garnish to salads.

Another fertilizer plant, chives accumulate potassium and calcium. I like to plant chives at the ends of my garden beds. Giving the plants a haircut a few times a year, it's easy to mulch the garden beds with the clippings for free fertilizer. The strong scent of chives is a deterrent to pests, so I plant it among my strawberry patch to deter pests attracted to the sweet scent of ripening strawberries. Pollinators will enjoy the beautiful flowers throughout late spring and early summer.

Chives are said to repel fruit tree borers and other fruit tree pests and diseases, so I planted a ring of chives around the trunks of each of my cherry trees. Chives are also said to be a good companion to carrots and tomatoes.

Chickweed *(Stellaria media)*

Chickweed shows up in disturbed soil such as garden beds and highly tilled areas, and is often considered a weed. Its presence can indicate low fertility. Chickweed accumulates potassium and phosphorus, two primary nutrients for healthy plant growth. Chickweed attracts pollinators searching for nectar in the spring and early summer, and also makes good poultry forage.

Chickweed will benefit the soil if left to grow and die back on its own. For a tidier garden, cut the plants back monthly (but leave the roots intact) and tuck the cut plant matter under the mulch. Or lay it on top of the soil to naturally decompose.

Comfrey *(Symphytum x uplandicum)*

Comfrey is the poster child for permaculture gardens, being perhaps the most important mulch plant. It's at the top of the list of natural fertilizers, accumulating potassium, phosphorus, calcium, and a handful of other nutrients in its large leaves. It is commonly planted underneath fruit trees and on the edges of the vegetable garden. Comfrey's large leaves can be chopped-and-dropped frequently throughout the season to feed the soil or to boost the compost pile. See chapter 3 for more information on using comfrey as a soil amendment.

The beautiful, bell-shaped purple flowers are popular with pollinators, and the giant leaves attract many types of beneficial insects looking for habitat.

Common comfrey (*Symphytum officinale*) is quick to self-seed, so I prefer to grow Russian comfrey, which has sterile seed and will play nicely in the garden without spreading.

Comfrey is also popular as poultry forage.

One of the most useful healing herbs, the dried leaves and roots of comfrey are often used in salves and tinctures.

Dandelion *(Taraxacum officinale)*

Dandelions—like many weeds—benefit our garden in many ways, the most important of which is fertilizer. Dandelions reach deep into the subsoil with those long taproots, dredge up important nutrients, and store them in their leaves. They excel at accumulating potassium, phosphorus, calcium, and a handful of other nutrients in their leaves, which are important for healthy plant growth. When those leaves die back or are cut back and left to decompose, they fertilize the soil.

I let dandelions grow in my vegetable garden, and it is common to encourage dandelions to grow in orchards under fruit trees. Dandelions increase earthworm populations, which is good for healthy soil. About once a month, I snip the leaves off and compost them in place, which also discourages the plant from flowering and going to seed. Dandelions are good, but I don't need a dandelion garden!

It is a favorite foraging plant for poultry.

Fennel *(Foeniculum vulgare)*

Fennel is a strong-scented plant with lacy foliage. The flower of fennel is umbel-shaped, like yarrow, and the beneficial insects and pollinators love it. Fennel attracts ladybugs, hoverflies, parasitic wasps, and lacewings, and I've enjoyed seeing an increase of swallowtail butterflies in my garden. I keep my fennel at the edge of my vegetable and fruit gardens because it's said that many

crops do not grow well with it, but I mulch with it at the end of the season after I've harvested all the seeds because it's a good accumulator of phosphorus.

TIP: HARVESTING FENNEL SEEDS

Harvest fennel seeds at the end of the summer so the plants don't set seed everywhere over the winter. Cook with the seeds, chew on them after meals to help with digestion, and give them away to your gardener friends for planting! If you have an out-of-control fennel patch, chickens will enjoy the forage.

Lamb's Quarters *(Chenopodium album)*

The presence of lamb's quarters is common in old farm fields, where chemical fertilizers were used in excess. Over time, these "weeds" will improve the soil quality. That's because lamb's quarters' deep roots accumulate nitrogen, phosphorus, potassium, calcium, and manganese while loosening compacted soil.

Lamb's quarters will benefit the soil if left to grow and die back on their own. However, one plant can set over 75,000 seeds. For a tidier garden, cut the plants back monthly to keep them from setting seed (but leave the roots intact). Tuck the cut plants under the mulch, or lay them on top of the soil to naturally decompose.

This is another super plant that is a favorite for poultry forage.

Lemon Balm *(Melissa officinalis)*

It wasn't until a few years ago that I learned the magic of this herb. As an accumulator of phosphorus, it is a wonderful herb to have growing in the vegetable garden or under fruit trees. It has a clumping growth habit, so it won't spread into areas you don't want it to, but it is fast-growing and prolific, so it can be cut back frequently to be used as a fertilizer.

Its scent will confuse pests in search of your vegetables or fruit crops. Often called the bee herb, the white flowers bloom all season and are popular with—you guessed it—bees. Lemon balm's foliage is a popular egg-laying site for lacewings, a beneficial insect.

With a lemony mint flavor, lemon balm has top-of-the line flavor as a culinary herb and as a tea. It has a long list of medicinal benefits to boot, and is commonly used in natural remedies.

Plantain *(Plantago spp.)*

Brought into North America by colonists, plantain often pops up where soil is compacted. Plantain accumulates calcium, sulfur, magnesium, manganese, iron, and silicon, and is therefore an excellent fertilizer plant to have around. Plantain will benefit the soil if left to grow and die back on its own. For a tidier garden, cut the plants back monthly (but leave the roots intact). Tuck the cut plants under the mulch, or lay them on top of the soil to naturally decompose.

This is another super plant that is a favorite as poultry forage.

White Clover *(Trifolium repens)*

One of the most useful additions to a productive garden, white clover is a nitrogen-fixing herb. All vegetable and fruit plants require nitrogen to produce healthy crops, but they can't access the nitrogen in the soil. Rather, they need certain nitrogen-fixing plants that take nitrogen from the air and convert it to a form that's useable by the roots of crops. That's where clover comes to the rescue.

There are many other nitrogen-fixing plants, but white clover is the easiest—in my opinion—to use in the garden. In addition to nitrogen, clover accumulates phosphorus. It is commonly used in orchards as a perennial living mulch under fruit trees to protect the soil and continuously provide fertilizer. It can also be grown in permanent garden pathways. It makes a great walkable ground cover, and the nitrogen will filter into the surrounding garden beds.

A favorite of honeybees, white clover will attract pollinators and beneficial insects because it blooms all season long. It's a common addition to poultry foraging mixes.

Yarrow *(Achillea millefolium)*

Yarrow is a gorgeous flower that is beloved by all manner of beneficial insects: ladybugs, hoverflies, parasitic wasps, and lacewings. I plant yarrow at the edge of the vegetable garden. Its scent will confuse pests trying to hone in on your vegetable crops!

Yarrow is also a great fertilizer, as its leaves are rich in potassium and phosphorus. When it finishes flowering in the fall, I chop it down and let it compost in place in the vegetable garden. It is also a nutrient-rich addition to the compost pile. Yarrow will fertilize soil and attract beneficial insects when planted under fruit trees.

Its deep roots will break up compacted soil or soak up extra water in a rain garden. It will also make a good ground cover when mowed.

While this is by no means an exhaustive list of super plants for the permaculture micro-farm, I hope this gives you some ideas of how you can use plants to improve your local ecosystem, increase the health of your garden, and reduce your workload.

Permaculture strategies can help us create functional, efficient, and low-maintenance micro-farms while building fertility and biodiversity, too. I hope you'll take advantage of some of these tools for using water wisely, using sloping land to your advantage, and using multi-functional plants in combinations that multiply yield and reduce work.

12 MAKING MONEY ON THE SUBURBAN MICRO-FARM

LAND IS ULTIMATELY THE MOST PRECIOUS RESOURCE WE HAVE.

ALTFRID KRUSENBAUM, ORGANIC VALLEY FARMER

THE SUBURBAN MICRO-FARM

Once you get going with your suburban micro-farm, you might wish that you could do it full time. If the only thing holding you back is money, that's good news! The sky's the limit as far as creative ideas for generating income from your micro-farm; it's merely a matter of finding a product or service to sell that matches your skills, interest, and resources.

But don't quit your day job just yet. It may take some time to gain momentum with your alternative income-generating ideas. One of the tricks to making income from the micro-farm is to diversify products and services. In other words, mix and match a few that meet your interest and skill level, and you'll be on your way to becoming a successful entrepreneur.

The following are examples of some ways to supplement your household income that don't require you to be a full-time market micro-farmer. This is by no means an exhaustive list. Remember, the sky's the limit!

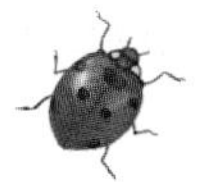

Perennial Crops

Perennial crops are an excellent addition to any income-generating micro-farm. Perennials come back every year without replanting, are only harvested once a year, and require little maintenance. Perennial crops aren't as demanding on the land as annuals, which will allow you to generate income while maintaining soil fertility. Also, perennials can improve the biodiversity of your property.

Some examples of perennial crops are asparagus, tree fruits, rhubarb, and even nuts. Asparagus is an excellent perennial crop because it is the first to produce in the spring, allowing you to start generating income earlier in the growing season. As a gourmet vegetable, both local consumers and chefs will pay top dollar for it. Shelf-stable, dried asparagus can be marketed in the off season or used to make soups and stews for home use.

> BY PLANTING A DIVERSITY OF CROPS WE HAVE HEDGED OUR BETS AND AREN'T PUTTING ALL OF OUR EGGS IN ONE BASKET. IF THE SEASON IS GOOD FOR ONE CROP, IT MIGHT NOT BE FOR ANOTHER.
>
> *MARK SHEPARD, RESTORATION AGRICULTURE*

When choosing tree crops, be aware of the harvest window of each. It will be important to stagger harvests so that you can stagger income throughout the growing season, as well as keep from being overwhelmed by harvesting and selling many crops at once. I found this out the hard way on my own micro-farm. My currants, cherries, black raspberries, and strawberries all come into harvest at relatively the same time. It made me feel overwhelmed—and this was only the fruit grown for household use! I've never seen a fruit vendor at a farmers' market without a line, so this tree fruit is definitely a smart choice.

Nut trees such as almonds, chestnuts, pecans, and walnuts—while slow to begin producing—will be profitable in the long-term. Additionally, all except the almond trees will produce high-value timber as a future income possibility. Hazelnut shrubs are also a possibility.

Strawberries and berry bushes are also a good choice, but harvesting from them will be more of a challenge than with other types of perennial crops. That's because the harvest window for these soft fruits doesn't last long, and it will be challenging to get them to market before the fruit starts to perish.

Asparagus likes moist areas where water collects briefly. You'll find wild asparagus growing naturally in ditches along a roadside.

Rhubarb is an easy spring crop to grow.

Many growers opt to host U-picks for their soft fruits, but you'll need a good insurance policy to cover all of the visitors. You'll also want to be located conveniently near customers. Promoting your U-pick will go a long way—potential customers won't understand the urgency of a harvest window without clear promotional efforts. Market farmers play an essential role in educating consumers who are used to getting the produce they want at any time of year in the grocery store.

Customers will need to be educated on when strawberries are ready for harvest, and how many weeks they can expect the harvest window to last. Otherwise, you'll have customers showing up at random times of the year asking about strawberries (This will inevitably happen anyway, even with good promotion). I remember many years ago when we helped start a farmers' market in our community. Customers showed up in May expecting to get tomatoes and corn, and were shocked and disappointed to find out that these were late summer crops.

Perennial herbs—both culinary and medicinal—are also great because they require almost no maintenance, are extremely prolific, and a little goes a long way. Most fresh and dried herbs command a good price at market, and demand is high. Herbs like cilantro, chives, and dill consistently command some of the highest profits per square foot, and are easy to grow.

Mark Shepard suggests that market farmers can reduce "not only farm costs, but family household costs as well." His family gets to eat the asparagus that isn't of a high enough quality to sell at market. They eat the "curled, undersized, oversized, too open, or insect-damaged asparagus spears that are perfectly nutritious and delicious to eat". What the family can't eat gets fed to the livestock, which then becomes nutrient rich, chemical free manure, keeping as many nutrients on the land as possible. *Restoration Agriculture* is chock-full of tips on how to grow perennial crops in an ecologically designed system to get a profitable return.

One of the main questions about selling a crop is where to sell it. Selling perennial harvests through farmers' markets, direct retail sale to customers who visit your farm, or wholesale distribution to natural foods markets or restaurants will require you to follow food safety regulations. One of the benefits of selling fresh, whole, uncut, unprocessed produce is that it will be exempt from health department inspections, food service permits, and licensing fees. Keep in mind that if crops are cut to offer samples—such as apple slices—a food safety permit and other sanitation steps would be required from your local health department.

The downfall of growing perennials is that they are more expensive to plant, and it will be at least a couple of years before they're ready to harvest. But perennials truly are a worthwhile investment. Planting them in stages can be the most cost-effective method for meeting short-term financial realities with the long-term economical benefit.

Annual Crops with a Long Season

Another smart solution for the suburban micro-farmer who can't be bothered with market crops that take all their time to manage, is to focus on annual crops that have a long growing season, only require a single harvest, and do not have strict storage requirements for taking them to market. Garlic, onions, potatoes, sweet potatoes, and winter squash (pumpkins, butternut, and acorn squash) are a few examples of plant-it-and-forget-it crops that travel easily to market with minimal damage or perishability.

Stay away from products like salad and leafy greens, which need a lot of soil enrichment, irrigation, and infrastructure to harvest and get to market before wilting. Bagged, cut leafy greens must be kept on ice in order to comply with most health department codes. I would also avoid tomatoes and peppers—unless you grow heirloom or unusual varieties—because everybody and their mother sells them, and they tend to go for pennies. As mentioned above, the fresh, unprocessed annual produce items that I've recommended in this section will sell in local markets without much regulatory oversight.

Garlic is likely unsurpassed in price per pound profits for an annual crop, and it will sell in almost any market—farmers' markets, retail shops, restaurants, or at your on-farm market. I like that it is planted in the fall, growing over the winter and early spring. When I'm busy getting the rest of my garden going in the early spring, I wouldn't need to be worried about planting my market crop.

Once the garlic is harvested, it may be the perfect time to plant sweet potatoes or winter squash for a late summer/fall harvest. That's right—the beds that grew garlic can be put into production for another market crop later in the same season.

While onions, potatoes and sweet potatoes can be purchased quite cheaply at the grocery store, discerning consumers and local chefs will look for unique, organic, and heirloom varieties grown in small batches, and are usually willing to pay a reasonable price. Potatoes can be planted in the spring for a summer harvest, and again in early summer for a late summer/fall harvest. Onions planted in spring will be ready for harvest mid-summer, and summer-planted onions will be ready for harvest in the fall. Pests generally do not bother sweet potatoes. They also have relatively low nutrient needs and will not be demanding on your soil.

One consideration is the annual investment of buying starts for garlic, onions, potatoes, and sweet potatoes if you aren't able to produce your own. This cost should be factored in when deciding which crops to grow.

Another consideration for these crops is harvesting time and method. Potatoes and sweet potatoes will probably require more time and effort to dig up than the other annual crops mentioned in this section. Winter squash can be harvested quite easily, while garlic and onions are visible from above and easily located.

Curing and Storing

Still, garlic, onions, potatoes, and sweet potatoes will need to be properly cured for about two weeks before taking them to market. This will ensure their tough skins will protect them against bumps and bruises in transport, and their flavors will be more developed as well. Be sure to let customers know if the curing hasn't been done and how they can cure their purchase at home to preserve the harvest.

When properly cured and stored, these annual crops can be sold throughout the fall and winter even while the garden rests, but this means that a storage area must be created to keep the produce in proper storage conditions.

Winter squash of all kinds are a celebration of the autumn season. Plant pie pumpkins or jack-o-lantern pumpkins for a popular fall market crop. Since they store well, you'll be able to sell them into the winter.

Shade Crops

Mushrooms, ramps, ginseng, goldenseal, and blue or black cohosh are examples of crops grown in the forest that have low maintenance requirements and command a high price at market.

Mushrooms

Mushrooms are probably the most popular crop of this bunch, but to sell them, you'll want to make sure to follow health department regulations. Wild-harvested mushrooms are generally not allowed to be sold due to the difficult nature of properly identifying various mushroom types and their safety for consumption.

However, because of the high price consumers and chefs are willing to pay for gourmet mushrooms, more and more cultivated mushroom operations are cropping up around metropolitan areas. Many of these interior setups use elaborate, sanitary procedures to comply with commercial regulations. However, the best mushrooms for micro-farmers are those that can easily be grown outside—without much oversight—on logs in the shade of a forested area or on the north side of a structure.

Garlic is a popular farmers' market product.

Pumpkins are a symbol of fall harvest celebrations.

Everybody loves farmer-grown potatoes!

Sweet potatoes are a reliable fall crop.

Shiitake, oyster, and lion's mane mushrooms are the most common varieties cultivated in small, log-grown operations. Reishi is a mushroom that is also grown on logs and is known for its potent medicinal qualities, popular at health food stores.

For all mushroom operations, there will be an initial investment to acquire the necessary materials. As with any of these income-generating methods, be aware of the setup costs and what you can expect to make from sales. According to a study by the University of Kentucky cooperative extension service, one person could manage an operation starting with 100 logs with an initial investment of around $400. This should give you an idea of what it may take to begin.

The harvest-to-market shelf life is very short for quickly perishing mushrooms, so be sure that you've spent some time developing a marketing plan and have found a buyer, or have already made arrangements for a farmers' market booth before your harvest is ready.

Dried mushrooms make an excellent holiday gift and this shelf-stable product can increase your sales throughout the winter. Mushroom powder is a nutritional product that can be added to soups and stews or can add flavor to sauces. Mushroom seasoning and soup blends are also popular choices for value-added products to bring to market.

Ramps grow in colonies.

Ramps are a member of the onion family and are native to North American hardwood forests. They are one of the earliest crops of the season and can help you begin generating income earlier in the year than you would with annual crops. High-end local chefs would be over-the-moon to find a sustainable, reliable source of ramps. That's because in recent times wild ramps have been over-harvested as this old-time favorite has fallen into gourmet popularity again. The over-harvesting has meant that long-standing groupings of ramps have begun to dwindle away. Consumers aren't as aware of what ramps are, but a little education and some ideas on how to use them (hint: like onions and garlic) will go a long way.

Ramps grow well on slopes that tend to stay shaded, cool, and moist year round. Typically

north-facing slopes are the most cool and moist. Rich soil and tree cover are important. They will improve the ecology of your forest (see chapter 11), but the drawback is that they will take five to seven years to be harvestable.

Ginseng, goldenseal, and blue and black cohosh are just a few of many medicinal herbs that naturally grow in deeply shaded forests. They can be wild-harvested or cultivated, and typically take around three to five years before they are ready for harvest. They can be sold as fresh roots, stems, bark, or leaves, but they are most often sold dried. It will be important to identify your market, whether it is a medicinal company searching for raw material or local herbalists. Additionally, you can consider creating tinctures, salves, soaps, or other value-added products, but these may need to follow FDA regulations.

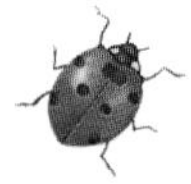 The Edible Plant Nursery

Many berry bushes, brambles, and herbs are easily propagated by cuttings from existing plants. They won't take up much room and will sell for a nice profit. This is a low-tech side job that doesn't require you to be an expert to get things going. Since you will be propagating from cuttings taken from existing plants, your startup costs will be low, and subsequently, so is risk. The trick is to get started soon, because plants will take about two to three years to be large enough to sell. If your plants that are large enough to sell do not sell right away in the first year, don't despair. They will continue to grow and the larger plant can be sold for an even higher price in following years.

FRUITS THAT PROPOGATE FROM CUTTINGS

- blueberry
- brambles (blackberry, raspberry)
- currant
- elderberry
- fig
- gooseberry
- grape
- hardy kiwi
- mulberry
- quince
- serviceberry

It might seem like you need a lot of space to start a plant nursery, but plants you intend to sell can be planted close together, since they don't require a permanent location. This will also help to cut down on weeds and aid in more efficient watering. Yes, one key component of a successful nursery business is that the plants are watered regularly in hot, dry weather.

Make sure you choose plant types and varieties that suit your climate and sun conditions. This is a good time to ask any potential customers (backyard gardeners) what kinds of plants they are looking for. Fruits that propagate from cuttings will be really popular.

Herbs are another easy nursery crop to propagate through cuttings. They are such a popular item for home growers that adding them to your nursery collection can really catapult plant sales. That's because herbs root and grow quickly into well-sized seedlings. No need to wait years to have a product for market!

Fruit trees are a little more complex to propagate, but once you get the hang of propagating the above plants from cuttings, you'll be ready to learn the process of fruit tree grafting. Then you'll be well on your way to having a value-packed micro-farm nursery business!

HERBS THAT PROPOGATE FROM CUTTINGS

- basil
- lavender
- lemon balm
- marjoram
- mint
- oregano
- rosemary
- sage
- thyme

Black raspberries started from cuttings

Oregano from cuttings

Cut Flowers

Growing a cut flower garden would not only be lovely and attract all kinds of wonderful pollinators and insects, but it would enable you to sell bouquets and pass on the cheer to others. Cut flower bouquets have a variety of opportunities for distribution. They can be sold at farmers' markets—it's where we got all of the flowers for our wedding—they can also be sold to natural food markets, restaurants, or direct to florists. Some businesses will even enjoy having cut flowers in their offices. You might even consider advertising to do flowers for events in your community.

There are even cut flower CSAs that have begun popping up in metropolitan areas. Imagine customers getting a weekly bouquet of flowers throughout the growing season to liven up their dining table, kitchen counter, or desk at work. Or, consider a cut flower U-Pick. Cut flowers can be an excellent addition to any one of these other income generation ideas. For example, imagine you have a booth at your farmers' market selling potted windowsill herbs, potatoes, mushrooms, and garlic. A colorful flower display will not only attract the eye of potential customers, but they might enjoy the additional option to bring home a bouquet.

Cut flowers will need plenty of sunshine, so if you have a shady yard, this option is likely not for you. Another challenge is matching flower types to your growing season. You'll want to choose flower varieties that have a long bloom time in your climate.

Flowers are usually sold by the bunch, bouquet, or individually. A mixture of perennial and annual flowers will help you to get started on the right foot, and it will ensure that you always have something blooming.

Most annual flowers will need to be planted frequently throughout the season to have a continuous supply of blooms.

QUICK-GROWING ANNUAL FLOWERS

- ageratum
- cosmos
- dahlia
- sunflower
- zinnia

Perennials are often used in cut flower arrangements because they are reliable. Know when each perennial will be in season and plan to have a variety of them staggered throughout the season for a continuous supply of predictable blooms. Mixing in common—but abundant—perennials like purple coneflower and black-eyed Susan can go a long way toward filling out bouquets.

To get started, identify a flat, cleared growing area in full sun. Till the soil and add soil amendments, or sheet mulch an area (see chapter 3 for details). Raised beds are becoming more popular with cut flower growers, as they make it easer to cultivate blocks of flower types together and harvest without fully bending over. Planning out an irrigation system beforehand will help your operation to be more successful. Additionally, mulching once your plants have reached 6 inches tall (see chapter 9), and cover cropping over the winter (see chapter 3) in annual beds will help you retain soil health and reduce weeds.

POPULAR PERENNIALS FOR BOUQUETS

- aster
- delphinium
- daisy
- iris
- liatris
- lily
- perennial sunflower
- rose
- tulip
- yarrow

Using a greenhouse or hoop house to start seeds earlier in the season can help you get your flowers to market sooner. Another way to capitalize on the short period of time you have to pro-

Daisies are prolific and easy to propagate.

Liatris is both beautiful and a favorite of pollinators.

duce flowers is to continuously grow annual flower starts so that you can quickly replace fading bloomers with new crops throughout the season.

When arranging a bouquet, stick to three to five flower varieties, and choose a two-color scheme—such as purple and yellow or purple and orange. The first of your three flower varieties will be the standout star of the bouquet. The second variety will have thick stems to provide structure for the bouquet, and the third variety will have an interesting texture to provide a backdrop for the others.

If you have an affinity for flowers, having a cut flower business may be right for you.

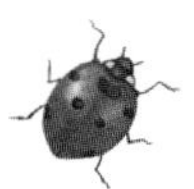

Maple Syrup

Maple syrup production is a labor-intensive process within a short season, but the once-per-year seasonality of it can make the profit worth it. Maples aren't the only trees that can be tapped. If you have birch or walnut trees, consider tapping them for their tree sap.

Regulations are the first thing to consider with regard to maple syrup production. Maple syrup is a value-added product, which means that it has been processed in some way—in this case, sap has been reduced to syrup. How the process is regulated from sap collection to the marketed product will depend on the department of health regulations of each state, so contacting your local office about possible permitting and inspections is a good idea. It could save a lot of headaches later. There are a number of regulations that specify the nature and sanitation of the sugar shack—the outbuilding where sap is boiled down to syrup. The final product in all instances must follow federal labeling guidelines. In some states, licensing requirements are exempt if gross sales are less than $15,000, even if the producer is still required to meet health and sanitation requirements.

Speaking of boiling sap into syrup: Did you know it takes 43 gallons of sap to make *one* gallon of maple syrup? Trees are tapped each year when the sap begins to flow, usually around mid-February when the days warm into the 40s and the nights are still freezing. Trees 10 inches in diameter and larger can be tapped. It is said that a stand of maple trees will be at least 40 years old before it is ready for tapping.

If you're not swayed by regulations and the sheer volume of sap required to make a gallon of syrup, then you'll want to focus on whether maple syrup production makes good financial sense. It's difficult to say what the profitability is because of the variations in the number of trees tapped, types of evaporating equipment, and the price per gallon in a local market. As with most markets, the more trees tapped, the better the profit. If you have a decent amount of land and can scale up your collection over time, then maple syrup production might be right for you.

Spend some time adding up the cost of equipment. You will have to acquire equipment for the collection of sap as well as equipment necessary for boiling the sap. The largest cost will be for the evaporator. Developing a process to streamline the collection of sap at a central location will go a long way toward reducing labor.

After the sap is collected, the next step is boiling the sap into syrup, which is the most time-intensive part. According to the Penn State University extension office, there are some sap collectors who choose to bypass this step. In some areas it is possible to sell your collected sap to a local maple syrup producer for $0.10-$0.70 per gallon and call it a day. If you enjoy the opportunity to be outside but can't invest in an evaporator, this might be a good option for you. Still, others enjoy the whole experience and will want to go the distance to become the syrup producer and market the final product.

Tapping birch or walnut trees for sap follows the same process for collecting and sugaring, but get this: you will need to harvest 100 gallons of birch sap to produce one gallon of syrup, but only around 16 gallons of walnut sap to produce one gallon of walnut syrup. With regard to production effort, then, it would seem like walnut syrup would be the way to go. This might be true, except that walnut trees produce far less sap than either maple or birch trees, so collecting 16 gallons of sap might not be as easy or quick. Additionally, the flavor profile is different in each. Maple flavor is mild and palatable, whereas birch syrup is more like molasses and walnut syrup is slightly bitter and not sweet.

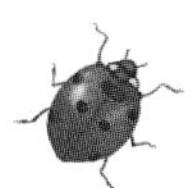

Cottage Food Products

There are a number of items that can be produced in the home kitchen and sold locally in most U.S. states: bakery products, jams and jellies, fruit butters, dried herb and seasoning blends, and dried tea blends, to name a few. Your state's department of agriculture will have guidelines to follow for the cottage food industry, but most cottage food facilities (read: home kitchens) are not subject to inspection or licensing. However, the department of agriculture is permitted to conduct random food samples to determine if the food product matches the label, so be sure your products are properly labeled.

The trick to selling cottage food products is promoting your product and establishing a customer base that is excited about what you are selling. Getting customer feedback and ideas on what they are willing and interested to buy is a good idea. Creative, attractive, and thoughtful packaging and a thoughtful display at your farmers' market booth will go a long way.

Cottage food products are generally only allowed to be sold in the state in which they were produced. If you live near state lines, this might limit your reach. However, they can be sold in any

food-based venues such as farmers' markets, natural food stores, or restaurants. Website sales for local delivery might also be an option to consider.

Bakery products are very popular cottage food products. Some examples of approved baked goods are cookies, breads, brownies, cakes, pies, no-bake cookies, unfilled baked donuts, waffle cones, and pizzelles. Though you may not be able to grow your own grains, you will want to create unique products that use on-farm ingredients such as chicken eggs, fruits, and herbs. On-farm ingredients will make your products stand out from the crowd.

Jams, jellies, and fruit butters are safe for home production because fruit has high acidity making the processed product a low risk as far as toxicity. Fruit products made from homegrown fruit are an economical product. While commercial producers have to purchase wholesale generic fruit, your supply of lovingly cultivated fruit is free! There is some labor involved in properly canning these items, but while the jars are in the canner, you can work on your awesome packaging that will attract customers.

Candy and sweet treats are popular, especially around the holidays. You can market them all year long by supplying treats for birthday parties and other special events. Try no-bake cookies, chocolate-covered pretzels, or similar chocolate-covered non-perishable items, such as chocolate-dipped granola bars.

Jams and jellies made from homegrown fruit is a delicious and economical product.

Busy moms and moms out shopping with their kids will appreciate your homemade snack products with honest ingredients. Popcorn, flavored popcorn, kettle corn, popcorn balls, caramel corn and nut snack mixes are some popular approved choices. Don't forget to figure out ways to include on-farm ingredients to give your products a creative edge in the market.

With all of the crap in packaged cereals today, you will go far if you can market a bulk-sized, healthy granola or dry cereal mix. Quick breakfasts with quality ingredients are a luxury that—if the right price point can be found—would make moms and busy, working folks really happy.

And finally, to capitalize on what you grow on your micro-farm, try marketing dry herbs, herb blends, dry seasoning blends, and dry tea blends. Products like these are marketable all year long to help extend income generation to the off-season months.

If you enjoy making delicious creations from your homegrown fruits, herbs, and even perhaps chicken eggs and honey, then a cottage food business may be right for you.

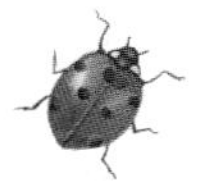

Homemade Soap

Homemade soaps are a lovely product that appeals to discerning consumers looking for products that are free of synthetic ingredients and made from pronounceable ingredients. Soaps are easy to sell at farmers' markets, natural food stores, on-site farm shop, or any boutique gift shop or bookstore. Holiday craft shows and festivals are good venues, too. Handmade soaps can also be sold online through online marketplaces like Etsy and your own website.

Before jumping in, browse some of these selling locations, both locally and online. Check out the product labels—the size of the bars, the wording on the labels, the ingredients used, and the prices charged. As you create bars that are uniquely yours—hopefully incorporating some on-farm ingredients like herbs or spices—you will have to assess the proper price point for your product that will make a profit. You'll have costs such as the purchase of your supplies and equipment, packaging, labeling, shipping materials and costs (if selling online), and marketing costs such as starting a website, logo and branding, etc. Additionally, there is insurance and taxes to consider. All of these costs will need to be accounted for when considering how to price your soap product. It may even be necessary to give samples away in order to attract customers. Don't forget to include your contact info on the labels!

Don't skimp on your marketing and labeling materials. If necessary, hire someone to help you with setting up a website, logo, and branding. Be sure to have brochures or business cards. Nowadays, people trust producers who have a website, where they can go to read about the product without feeling pressured at a market booth. You will also have to refine your selling personality. Few people feel confident about sales, but it is a necessary part of marketing a product. Practice

speaking confidently and enthusiastically—yet with a friendly and non-pressuring voice—about your product.

To avoid regulatory nightmares, do NOT make any claims on your labeling that your product treats or prevents disease, or affects health in any way. Do not claim, for example that it moisturizes skin, makes the user smell nice, or deodorizes. This will make your soap a cosmetic, requiring FDA regulation. Any color additives will also require FDA approval and oversight, so I would avoid those, too. Any medicinal claims will make your product a drug, again, requiring FDA approval.

If you really enjoy making soap, you might want to expand your product line and start making other body care products such as salves, lotions, lip balm, skin creams, and shampoos. This is entirely possible, but keep in mind that these items are considered cosmetics, and are therefore subject to regulation. Your products might need to be made in an approved, commercial grade kitchen and be reviewed by food safety inspectors.

Be sure to test your products before selling them. This seems obvious, but it needs to be said, to ensure that your products are safe and effective. Get feedback from family and friends before taking your product to market. Seasonal products during the holidays will catapult your end-of-year sales. Make sure to include products for men and kids, too. Soaps make great party or wedding favors, so be sure to market a size and price point that works for larger orders.

Soap makers are highly motivated producers that have to get creative about how to make their products unique and how to turn a profit from them.

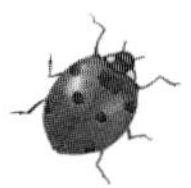

Teaching and Writing

My tiny yard wasn't conducive to producing excess fruits and vegetables for sale beyond feeding our household, so market farming hasn't been a focus for me. Rather, I've focused on education as a source of income through my blog at TenthAcreFarm.com and writing projects such as articles for online magazines. I occasionally teach classes and offer permaculture design consultation locally.

If education—rather than a physical product—is more your style, here are some tips for utilizing your talents and making a return on investment.

Teach what you know.

If you have a skill that people want to learn, chances are you have an opportunity to teach classes or workshops—either locally or digitally; or speak to groups. I have helped to teach local classes on topics such as eating locally, installing rain barrels, rain catchment, compost systems, and terraces, as well as topics related to permaculture gardening. I've had the opportunity to speak

to local garden clubs about community gardening and edible landscaping, and I've been able to speak to conservation groups about what I call *edible restoration*.

Give tours or offer consulting.

If you have a well-established garden, consider giving tours along with mini-classes. Some of the most popular classes I've ever taught were those taught at my home in conjunction with a garden tour. People love to tour others' gardens.

Tours are great to stack with other products you might have to sell. While people are visiting your micro-farm, they would love the opportunity to buy products that came directly from your own land and were processed with your own hands. In a world where everything was made on a faraway assembly line, consumers appreciate being able to see with their own eyes where things they buy come from, the hands that grew and harvested it, and the person that crafted it into a final product. People crave products that have a story and come from a special place that is well cared for and loved.

Or you could become a garden consultant, helping others to establish and maintain backyard gardens. More and more people want to start their own micro-farms but have no idea where to

The Permaculture Yard Class & Tour @ Tenth Acre Farm (photo by Ken Stigler Photography)

start. A garden consultant is very much an educator, as they guide the homeowner toward a plan that helps them take actionable steps toward a productive homestead.

Run a blog or write for online outlets.

A blog is a great way to get to know your audience and establish your authority, credibility, and integrity at your skill. Writing high-quality content related to your field of expertise is the key to engaging existing—and attracting new—customers/students. Whether your product is a book that you wrote or apples from your orchard, a blog—with an accompanying newsletter—provides valuable information for your customers and keeps them up to date on what you have going on. People like the opportunity to get to know the trusted face behind their products and Google searches!

TenthAcreFarm.com

If you are an expert with a unique farm-y skill and also have writing skills, consider writing for online outlets, such as magazine websites. Increasingly, many of these outlets want to work with writers who already maintain a blog.

Keep in mind that generating an income from teaching and writing is not something that usually happens overnight. I spent the better part of two years teaching and writing for free before I was able to land paid gigs. There is a learning curve, and it takes a while to become a good speaker and communicate your expertise well. However, if you like talking to and relating to gardeners and homesteaders, then teaching, consulting, and/or writing might be a good way for you to generate supplemental income for your family while doing what you love.

Tips for Success

While there are a lot of exciting opportunities for making money from your micro-farm, your success will ultimately come down to how well you manage the business and marketing side of things. Here are a few tips that can help you successfully market your product.

Establish a Product Distribution Method

Depending on where you live, distribution of your product in your local community could be a challenge. Think about who your customers would be and how you would get the word out to them. Would you sell at a farmers' market? To restaurants? Here in Cincinnati, we have a Yahoo group with over a thousand 'locavore' consumers and chefs, and it's where micro-farmers can advertise when they have a harvest or product available for sale.

Some growers set up their backyards like U-picks when a crop is ready for picking, but be sure to check on insurance issues before doing something like that.

As far as the distribution of education or non-food items, taking advantage of the Internet allows you to have a global customer base.

Build a Customer Base, Develop Relationships

The best way to have your hard work compensated properly is to develop a public image that will be seen by your potential customers. If your customers are residents in your neighborhood, become a good neighbor (Not just to sell products, but also because it's more fun!). Be social, friendly, and thoughtful. Participate in or host neighborhood events, like a block party. Set out free samples, give tours, and hold workshops.

Talk about your product and poll your potential customers for ideas. If half of them would buy blueberry bushes but only one would buy a kiwi vine, then focus on propagating blueberries.

Whether your customers are in your neighborhood or at the farmers' market, being social will go a long way. I'm talking social media. Keep a Facebook page, a Twitter account, and/or an Instagram account for your business so you can let customers know when new products are available and how they can get them. Have a website or blog so potential customers can read about your business before buying. Connecting the website to a newsletter will go a long way to legitimize your business and keep customers informed of your progress and your offerings.

Tell stories and share your journey so customers can appreciate the care you put into developing quality products.

Regulations and Legal Mumbo-Jumbo

Keep records of your profit and loss over time, so you can make the most out of tax season. Internal Revenue Service Code 183 helps distinguish between a hobby and a for-profit venture and will help you figure out if you are eligible for deductions.

Always check local regulations before starting a business. Some regulations will cover the property itself for home-based business, while others will cover restrictions on the products you're interested in selling. Consider talking to a lawyer about the appropriate business entity for you (LLC, sole proprietorship, etc.), and whether you need to carry liability insurance. Doing your homework beforehand will ensure that you don't waste time or money.

In the end, managing your land ecologically and reaping edible abundance from it is the gateway into wanting to spend all your time at home! Luckily, there are many creative ways we can go about bringing extra income to the micro-farm. With just the right combination of products and entrepreneurial spirit, we may even be able to become full-time micro-farmers.

THE ULTIMATE GOAL OF FARMING
IS NOT THE GROWING OF CROPS,
BUT THE CULTIVATION AND
PERFECTION OF HUMAN BEINGS.

MASANOBU FUKUOKA,
THE ONE-STRAW REVOLUTION

CONCLUSION

Micro-farmers are at the forefront of writing a new story about how suburbanites engage with their environment. As once-fertile farmland is transformed into housing developments at an ever-faster rate, it is essential that we learn efficient ways of growing food that protect the dwindling rural and wild lands that remain. Saving them from the grip of industrial farming and its ecosystem-stripping practices is essential. Though we have long depended on far away lands to supply our produce, many of us now aim to grow some of our own healthy produce right where we live, using the suburbs' most valuable asset: the under-utilized lawn.

The compact suburban micro-farm may look different from the traditional farm. It can either conform to the existing landscape or make a bold statement depending on the needs and desires of the micro-farming homeowner. No matter your farming style, one thing is for sure: The suburban micro-farm has been shown to be more productive than farmland, so it is neither a waste of time nor space.

Modern life is busy and our yards may not be perfect farming sites. Still, it isn't necessary to have a lot of time or flat, sunny space to be a successful micro-farmer. All we need is a plan and a willingness to seek out micro-farming practices that match our schedule and improve efficiency. Producing chemical-free food in lieu of resource-intensive lawn may deliver cost savings both at the grocery store and at the pharmacy. As the number of professional farmers declines, our micro-farming skills and productivity can contribute to overall food security.

We now realize that a productive micro-farm doesn't have to come at the expense of our local ecosystem. Rather, with intentional action, biodiversity and soil fertility can skyrocket. Strong, ecologically friendly gardens will be better able to respond to drought, flood, chemical invasions, and pests.

This won't happen overnight, however. Creating an interconnected ecosystem will take time as well as a certain amount of trial and error, as it is a more complex affair than simply building a garden. Patience will be required as we balance modern life with building a resilient, productive micro-farm. It will take work, but we will certainly be rewarded. I am more convinced than ever that a committed group of micro-farming pioneers can positively affect the health of both our inhabited and wild lands.

Use the tools in this book, first and foremost, to help you keep records, manage your time, and build the soil; but refrain from trying to master every strategy before beginning to build your micro-farm. The details can be learned over time. Indeed, you will learn as you go and discover nuances that will be unique to your environment and your situation. You may not be able to solve all the world's problems on your own micro-farm, but you can certainly become a steward of your little piece of heaven. Many people in the world can only dream of this opportunity.

ACKNOWLEDGEMENTS

To say that I'm shocked to call myself an author would be an understatement. Life has been a winding road, and many wonderful humans have supported me along my journey to this place, where I get to share my passion of growing edible things with the world in words.

To Vince, my companion and most faithful cheerleader, who never wavered in his confidence that I had something to offer the world: You gave me the space and the tools to explore this path, and the nudging I needed to turn it into something meaningful. This book is yours as much as mine.

To my parents, who gifted me with a tenacious stubbornness to see any project through to completion. To Tim, Ben, Karyn, Lisa, Barb, and Lynn, who knew me before I knew myself and didn't balk (out loud) at my notion to farm my yard and then tell the world about it.

To Sharon, who taught me to love growing things, and to Jane and Randy, who gave me an opportunity to explore my writing voice.

To Braden and the Cincinnati permaculture community: Your presence has helped shape my path, and to John and Madeline, whose friendship began in that place. You gave this book a boost when I needed it most.

To the gardeners of Hillside Community Garden: Barb, Maria, Rose, Tim, Ron, and Patsy: Thanks for believing in me. And to Peter, Bill, and Winnie for supporting this project that enriched my understanding of the intersection between food production and ecology.

To Suellyn for encouraging me to keep writing, and all my friends of the Enright Ecovillage, whose many demonstrations of yard farming made an imprint on my mind and motivated me to continue working to supplant the suburban lawn.

Also thanks to my fellow local suburban homesteaders: Julie C and Mark, Julie R, Melanie, and Karen, and those afar on the interwebs—who shine like beacons and assure me that there is a place for me in this world.

To the Supernaturalists—especially Beth, Bob, Cam, and John, who have supported my journey carte blanche. To Ed and Julie, who never raised an eyebrow when we dug up our front yard.

To the readers of TenthAcreFarm.com: Your amazing loyalty and feedback is what encouraged me to put my words into long form.

To Julie, Brannan, Becky, Wendy, Melinda, and countless others who mentored me through book publishing: Thank you for your patience and for improving this work far beyond my expectations.

To Molly the cat, for being a mischievous writing companion and a distraction from my thoughts. Thanks, I guess.

And finally, to all of you who will find inspiration from this book, thank you for reading. May we all work together to take responsibility for our existence on Earth.

BIBLIOGRAPHY

5 Most Profitable Nut Trees to Grow
http://www.profitableplants.com/5-most-profitable-nut-trees-to-grow/

2015 Gulf of Mexico Dead Zone 'Above Average'
http://www.noaanews.noaa.gov/stories2015/080415-gulf-of-mexico-dead-zone-above-average.html

2015 Shopper's Guide to Pesticides in Produce
http://www.ewg.org/foodnews/summary.php

Ag 101 "Demographics"
http://www.epa.gov/agriculture/ag101/demographics.html

All New Square Foot Gardening: The Revolutionary Way to Grow More in Less Space
by Mel Bartholomew

Americans in Debt
http://www.debt.org/faqs/americans-in-debt/

Building Healthy Soils in Vegetable Gardens: Cover Crops Have Got it Covered Part 1: Introduction to Cover Cropping
http://blogs.extension.org/gardenprofessors/2015/03/19/building-healthy-soils-in-vegetable-gardens-cover-crops-have-got-it-covered-part-i-introduction-to-cover-cropping/

Can Cities Become Self-Reliant in Food?
http://www.sciencedirect.com/science/article/pii/S0264275111000692

Carrots Love Tomatoes: Secrets of Companion Planting for Successful Gardening
by Louise Riotte

Coffee Grounds Perk Up Compost Pile with Nitrogen
http://oregonstate.edu/ua/ncs/archives/2008/jul/coffee-grounds-perk-compost-pile-nitrogen

Cottage Food Production Operation
http://www.agri.ohio.gov/foodsafety/docs/CottageFoodOperation-factsheet.pdf

Crop Profile: Asparagus
http://www.hobbyfarms.com/crops-and-gardening/growing-asparagus-14845.aspx

Crushed Eggshells in the Soil
http://www.aces.edu/timelyinfo/Ag%20Soil/2005/November/s-05-05.pdf

Cut Flower Production
http://extension.psu.edu/business/ag-alternatives/horticulture/cut-flower-production

Demographic Trends in the 20th Century
http://www.census.gov/prod/2002pubs/censr-4.pdf

Does Your Lawn or Garden Need Lime?
http://pender.ces.ncsu.edu/2013/01/does-your-lawn-or-garden-need-lime/

Edible Forest Gardens, Volume 2: Ecological Design and Practice for Temperate-Climate Permaculture
by Dave Jacke and Eric Toensmeier

Edible Landscaping
by Rosalind Creasy

Environmental Benefits of Manure Application
http://www.extension.org/pages/14879/environmental-benefits-of-manure-application#.VZ8mHpNVhBf

Food Not Lawns: How to Turn Your Yard into a Garden and Your Neighborhood into a Community
by H.C. Flores

Forest Farming Ramps
http://nac.unl.edu/documents/agroforestrynotes/an47ff08.pdf

Gaia's Garden: A Guide to Home-Scale Permaculture, Second Edition
by Toby Hemenway

The Garden Controversy: A Critical Analysis of the Evidence and Arguments Relating to the Production of Food from Gardens and Farmland
http://vufind.carli.illinois.edu/vf-uic/Record/uic_1190545/Description

Guidance on Maple Syrup Production
http://www.in.gov/isdh/files/maple_syrup_guidance_final.pdf

Guide for Organic Crop Producers
http://www.ams.usda.gov/AMSv1.0/getfile?dDocName=STELPRDC5101542

Greensand as a Soil Amendment
http://ucanr.org/sites/nm/files/76652.pdf

Growing Forest Botanicals and Medicinals
http://www2.ca.uky.edu/agc/pubs/FOR/FOR91/FOR91.pdf

Growing Mushrooms Commercially
https://www.mushroompeople.com/growing-mushrooms-commercially/

Herbicide Carryover in Hay, Manure, Compost, and Grass Clippings
http://content.ces.ncsu.edu/herbicide-carryover.pdf

Herbicide Contaminants in Purchased Straw, Compost, Manure, and Dairy Waste
http://sustainableneseattle.ning.com/profiles/blogs/herbicide-contaminants-in

Herbicides in Compost
http://whatcom.wsu.edu/ag/aminopyralid/CompostConcerns2010.pdf

Home Fruit Growing—Making More Plants
http://pss.uvm.edu/homefruit/hfgprop.htm

How to Farm Your Parking Strip
http://www.houzz.com/ideabooks/24109664/list/How-to-Farm-Your-Parking-Strip

How to Grow More Vegetables (and Fruits, Nuts, Berries, Grains, and Other Crops) Than You Ever Thought Possible on Less Land Than You Can Imagine
by John Jeavons

How to Make $40,000 Growing Garlic
http://www.profitableplants.com/garli/

Illinois Farmers Market Food Safety
http://www.idph.state.il.us/about/fdd/ILFarmersMrktFoodSafety.pdf

In the Know: U.S. Berry Demand and Imports
http://theproducenews.com/markets-and-trends/8784-in-the-know-u-s-berry-demand-and-imports

Killer Compost Reports: Contaminated Manure and Herbicide Contamination Damaging Gardens
http://www.motherearthnews.com/homesteading-and-livestock/sustainable-farming/killer-compost-herbicide-contamination-zl0z1211zkin.aspx

Landscaping with Fruit
by Lee Reich

Leaf Mulch Forum: "Research and Real-World Techniques"
http://archive.lib.msu.edu/tic/mitgc/article/199866b.pdf

Looking for Lawns
http://earthobservatory.nasa.gov/Features/Lawn/printall.php

The Lucrative Sweet Potato Takes Root
http://www.sare.org/Learning-Center/From-the-Field/Southern-SARE-From-the-Field/The-Lucrative-Sweet-Potato-Takes-Root

Making Money with Strawberries
http://www.hobbyfarms.com/farm-marketing-and-management/make-money-with-strawberries.aspx

Managing Cover Crops Profitably
by SARE (Sustainable Agriculture Research & Education)

Maple Syrup Production
http://extension.psu.edu/business/ag-alternatives/forestry/maple-syrup-production

Massive Scam Threatens Your Health—by Placing Toxic Chemicals on Land, Polluting Industries Are Allowed to Bypass Clean Air and Water Regulations
http://articles.mercola.com/sites/articles/archive/2015/11/01/biosolids-fertilizer.aspx

Maximilian Sunflower Plant Guide
http://plants.usda.gov/plantguide/pdf/pg_hema2.pdf

Most Profitable Fruits to Grow in Your Home Garden
http://www.cheapvegetablegardener.com/most-profitable-fruits-to-grow-in-your-home-garden-2/

The Most Profitable Plants in Your Vegetable Garden
http://www.cheapvegetablegardener.com/most-profitable-plants-in-your/

Mushroom Farming
http://www.hobbyfarms.com/crops-and-gardening/mushroom-farming-14815.aspx

Only 60 Years of Farming Left If Soil Degradation Continues
http://www.scientificamerican.com/article/only-60-years-of-farming-left-if-soil-degradation-continues/

Organic Coffee: Why Latin America's Farmers Are Abandoning It
http://www.csmonitor.com/World/Americas/2010/0103/Organic-coffee-Why-Latin-America-s-farmers-are-abandoning-it

Organic Mulching Materials for Weed Management
http://www.extension.org/pages/65025/organic-mulching-materials-for-weed-management#.VZ8pt5NVhBe

Oxford American College Dictionary
by Oxford University Press

Pests: Organic Garden Basics Volume 7
by the Editors of Organic Gardening Magazine

Plant Nutrients
http://www.ncagr.gov/cyber/kidswrld/plant/nutrient.htm

Potential Profits from a Small-Scale Shiitake Enterprise
http://www2.ca.uky.edu/agc/pubs/for/for88/for88.pdf

Restoration Agriculture: Real-World Permaculture for Farmers
by Mark Shepard

Seed to Seed: Seed Saving and Growing Techniques for Vegetable Gardeners
by Suzanne Ashworth

Selling Handmade Products (FAQs Series)
http://thenerdyfarmwife.com/selling-handmade-products-faqs-series/

Soap: FAQs
http://www.fda.gov/Cosmetics/ProductsIngredients/Products/ucm115449.htm

Soil Contaminants and Best Practices for Healthy Gardens
http://cwmi.css.cornell.edu/Soil_Contaminants.pdf

Steps Toward a Successful Transition to No-Till
http://extension.psu.edu/publications/uc192

Stern's Introductory Plant Biology
by James Bidlack and Shelley Jansky

Summary Report, 2010 National Resources Inventory
http://www.nrcs.usda.gov/Internet/FSE_DOCUMENTS/stelprdb1167354.pdf

Superbia! 31 Ways to Create Sustainable Neighborhoods
by Dan Chiras and Dave Wann

Using Manure and Compost as Nutrient Sources for Fruit and Vegetable Crops
http://www.extension.umn.edu/garden/fruit-vegetable/using-manure-and-compost/

Using Organic Matter in the Garden
http://www.gardening.cornell.edu/factsheets/orgmatter/

Using Wood Chips in a Vegetable Garden
http://learningandyearning.com/wood-chips

What Is Mushroom Compost?
http://extension.oregonstate.edu/gardening/what-mushroom-compost

The Winter Harvest Handbook: Year-Round Vegetable Production Using Deep-Organic Techniques and Unheated Greenhouses
by Eliot Coleman

Wood Ash Can Be Useful in the Yard If Used with Caution
http://extension.oregonstate.edu/gardening/wood-ash-can-be-useful-yard-if-used-caution

ADDITIONAL RESOURCES

Please visit my website, www.TenthAcreFarm.com, for my expanding collection of micro-farming resources. For the free supplemental material that accompanies this book, go to www.TenthAcreFarm.com/tsmf-companion. You'll find downloadable garden planning resources and additional resource links.

BOOKS

Here are a number of books that have influenced me as I seek to fine-tune my skills in micro-farming, edible gardening, permaculture, and other related topics. The following resources are listed in alphabetical order.

Animal, Vegetable, Miracle
by Barbara Kingsolver

Backyard Farming on an Acre (More or Less)
by Angela England

The Backyard Homestead: Produce all the Food You Need on just a Quarter Acre!
Edited by Carleen Madigan

Bioshelter Market Garden: A Permaculture Farm
by Darrell Frey

The Complete Idiot's Guide to Composting
by Chris McLaughlin

DIY Projects for the Self-Sufficient Homeowner: 25 Ways to Build a Self-Reliant Lifestyle
by Betsy Matheson

Farming the Woods: An Integrated Permaculture Approach to Growing Food and Medicinals in Temperate Forests
by Ken Mudge and Steve Gabriel

Gardening Without Work: For the Aging, the Busy & the Indolent
by Ruth Stout

Homegrown Herbs: A Complete Guide to Growing, Using, and Enjoying More than 100 Herbs
by Tammi Hartung

Making It: Radical Home Ec for a Post-Consumer World
by Kelly Coyne and Erik Knutzen

The One-Straw Revolution: An Introduction to Natural Farming
by Masanobu Fukuoka

The Permaculture Handbook: Garden Farming for Town and Country
by Peter Bane

The Resilient Farm and Homestead: An Innovative Permaculture & Whole Systems Design Approach
by Ben Falk

The Self Sufficient-ish Bible: An Eco-Living Guide for the 21st Century
by Andy and Dave Hamilton

Soil Biology Primer
by Elaine Ingham

The Urban Homestead: Your Guide to Self-Sufficient Living in the Heart of the City
by Kelly Coyne and Erik Knutzen

Weedless Gardening
by Lee Reich

ONLINE RESOURCES

The world is at our fingertips now! The following websites can help you quickly find useful information about micro-farming topics.

The 104 Homestead **http://www.104homestead.com/**

Attainable Sustainable **http://www.attainable-sustainable.net/**

Common Sense Homesteading **http://commonsensehome.com/**

Grow a Good Life **http://growagoodlife.com/**

Grow Forage Cook Ferment **http://growforagecookferment.com**

The Herbal Academy **http://www.theherbalacademy.com/**

Hobby Farms **http://www.hobbyfarms.com**

Homespun Seasonal Living **http://homespunseasonalliving.com/**

Homestead Honey **http://homestead-honey.com/**

Homestead Lady **http://www.homesteadlady.com/**

Joybilee Farm **http://joybileefarm.com/**

Kitchen Gardeners International **http://kgi.org**

Learning and Yearning **http://www.learningandyearning.com**

Root Simple **http://www.rootsimple.com/**

INDEX

Page numbers appearing in italic type refer to pages that contain illustrations. Page numbers followed by the letter "t" refer to tables.

Thank you for reading *The Suburban Micro-Farm!*

Please share your thoughts!

Consider submitting a review on the website of the bookstore where you purchased this book, or your favorite online bookstore.

Share your reactions on Goodreads, Facebook, or Twitter, and be sure to tag **@TenthAcreFarm**!

Sharing your thoughts with other people about this book goes a long way to support the author, and provides priceless feedback for future editions and/or new books!

Are you ready to grow food and re-wild your landscape using permaculture?

Start your adventures at TenthAcreFarm.com.

Also, don't forget to download your free bonus materials:
TenthAcreFarm.com/tsmf-companion

ABOUT THE AUTHOR

Amy Stross is a permaculture gardener, writer, and educator with a varied background in home-scale food production.

As a professional gardener and permaculture designer, Amy has specialized in creating ecologically regenerative and edible landscapes. Her own 0.10-acre micro-farm is a thriving example, which includes berry bushes, fruit trees, herbs, flowers, and vegetable gardens. She also ran a 5-year community food forest project at her local university, featuring a variety of edible gardens.

Her current adventure is transforming a 3-acre property into a micro-farm with her husband and mischievous farm cat. She reaches hundreds of thousands of people with her adventures and expertise in small-scale permaculture gardening on her popular website, TenthAcreFarm.com.

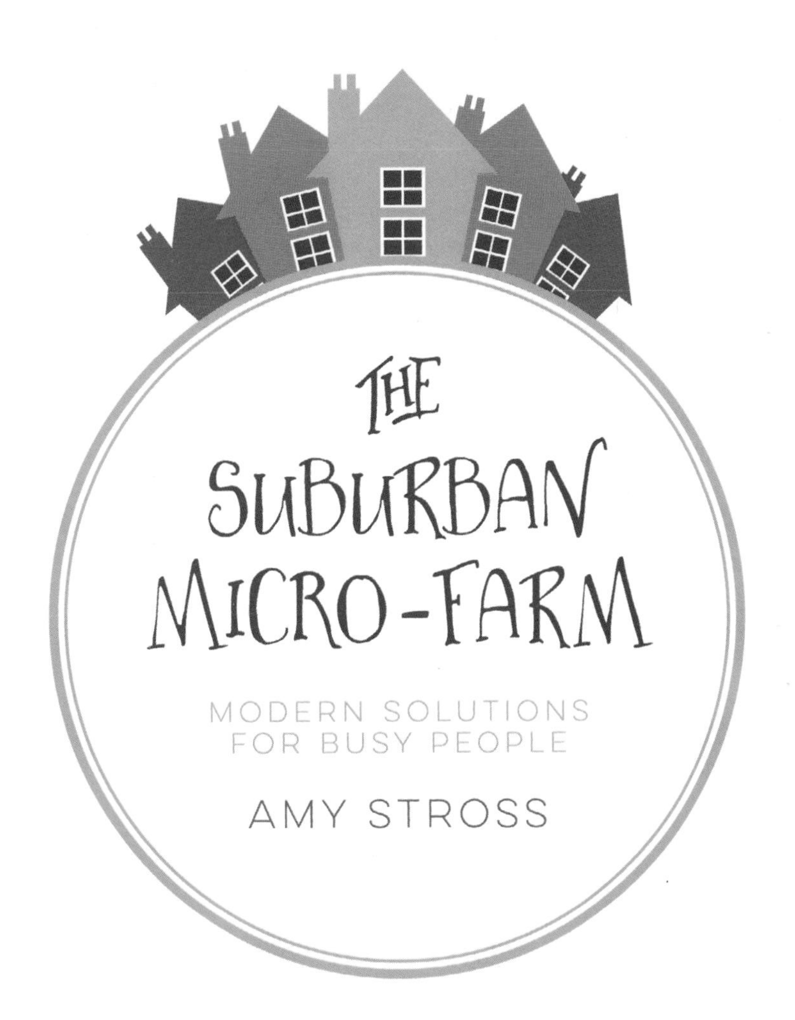

THE SUBURBAN MICRO-FARM

MODERN SOLUTIONS FOR BUSY PEOPLE

AMY STROSS

TENTHACREFARM.COM